Anticorrosion Protection of Nonmetallic and Metallic Coatings

Anticorrosion Protection of Nonmetallic and Metallic Coatings

Guest Editors

Tadeja Kosec
Aleksandra Kocijan

MDPI

Basel • Beijing • Wuhan • Barcelona • Belgrade • Novi Sad • Cluj • Manchester

Guest Editors
Tadeja Kosec
Laboratory for Metals, Corrosion and Anticorrosion Protection
Slovenian National Building and Civil Engineering Institute
Ljubljana
Slovenia

Aleksandra Kocijan
Laboratory of Analytical Chemistry and Corrosion
Institute of Metals and Technology
Ljubljana
Slovenia

Editorial Office
MDPI AG
Grosspeteranlage 5
4052 Basel, Switzerland

This is a reprint of the Special Issue, published open access by the journal *Coatings* (ISSN 2079-6412), freely accessible at: https://www.mdpi.com/journal/coatings/special_issues/Anticorrosion_Coating.

For citation purposes, cite each article independently as indicated on the article page online and as indicated below:

Lastname, A.A.; Lastname, B.B. Article Title. *Journal Name* **Year**, *Volume Number*, Page Range.

ISBN 978-3-7258-4655-9 (Hbk)
ISBN 978-3-7258-4656-6 (PDF)
https://doi.org/10.3390/books978-3-7258-4656-6

Contents

About the Editors

Tadeja Kosec

Tadeja Kosec is a senior researcher and Head of Laboratory for metals and corrosion and anticorrosion protection at the Slovenian National Building and Civil Engineering Institute (ZAG), with a PhD in Chemistry. Her work focuses on corrosion science, protective coatings, and the long-term durability of metallic materials in extreme environments. She has led national and international research projects related to infrastructure protection, radioactive waste disposal, and bronze conservation.

Kosec actively collaborates with various international institutions. She has authored numerous peer-reviewed articles and regularly presents at international conferences. Recognized for her mentorship and leadership, she has supervised doctoral students and contributed to science policy development, including funding strategy initiatives in Slovenia.

Her work bridges theoretical studies and practical applications, with emphasis on corrosion mechanisms, microscopy, and inhibitor development. She continues to promote excellence in research, heritage preservation, and sustainable materials science.

Aleksandra Kocijan

Aleksandra Kocijan is a Head of Laboratory of Analytical Chemistry and Corrosion at Institute of Metals and Technology (IMT), Ljubljana, Slovenia. She is a President of Scientific Council at IMT, Member of Editorial Board of the journal *Materials and Technology* and Leader of Infrastructure program at IMT. Her fields of expertise are the physical chemistry of metallic surfaces, corrosion, metallic material characterization, and biocompatible and bioresorbable metallic materials. She has experience in the fields of electrochemistry, analytical chemistry and surface analysis. Her work is dedicated to the exploration of the corrosion and passivation processes of modern metallic materials under various experimental environments, with special emphasis on mimicking biological and physiological conditions. She is an author or co-author of numerous original scientific papers, conference contributions, invited lectures, national and international projects, has participated in national programs and has H-index of 24. She is a mentor or co-mentor to several PhD students and, through her membership at the Scientific Council, she actively participates in establishing new research guidelines for the institute in accordance with national and international priorities.

Preface

The Reprint "Anticorrosion Protection of Nonmetallic and Metallic Coatings" compiles cutting-edge research focused on advanced corrosion-resistant coatings for a wide range of industrial and environmental applications. The present Reprint focuses on developments in metallic, nonmetallic, and composite coatings using different engineering approaches to increase their durability in various demanding applications. Key areas covered include the application and analysis of coatings engineered through different approaches such as physical and chemical vapor deposition, atomic layer deposition, thermal and plasma spraying, directed energy techniques, wet chemical and electrochemical processes, and techniques based on nanotechnology. The research featured in this Special Issue Reprint provides comprehensive insights into corrosion resistance, mechanical properties, and surface and interface characterisation, with emphasis on enhanced functional performance for different applications. This Reprint serves as a current reflection of the evolving strategies in anticorrosion technology, addressing both fundamental mechanisms and applied solutions for surface protection in real-world systems.

Tadeja Kosec and Aleksandra Kocijan
Guest Editors

coatings

MDPI

Article

Cyclic Hot Corrosion Failure Behaviors of EB-PVD TBC Systems in the Presence of Sulfate and Vanadate Molten Salts

Yasin Ozgurluk [1,*], Kadir Mert Doleker [1], Dervis Ozkan [2], Hayrettin Ahlatci [3] and Abdullah Cahit Karaoglanli [1]

1 Department of Metallurgical and Materials Engineering, Faculty of Engineering, Bartin University, 74100 Bartin, Turkey; mertdoleker@gmail.com (K.M.D.); cahitkaraoglanli@gmail.com (A.C.K.)
2 Department of Mechanical Engineering, Faculty of Engineering, Bartin University, 74100 Bartin, Turkey; dervisozkan@gmail.com
3 Department of Metallurgical and Materials Engineering, Faculty of Engineering, Karabuk University, 78050 Karabuk, Turkey; hahlatci@karabuk.edu.tr
* Correspondence: ozgurlukyasin@gmail.com

Received: 28 January 2019; Accepted: 20 February 2019; Published: 1 March 2019

Abstract: The cold gas dynamic spray (CGDS) method has been considered a promising technology to produce a metallic bond coat for thermal barrier coating (TBC) systems. In this study, CoNiCrAlY bond coats produced by CGDS method were coated with yttria-stabilized zirconia (YSZ) by electron beam physical vapor deposition (EB-PVD). TBCs were subjected to 50 wt % V_2O_5 and 50 wt % Na_2SO_4 molten hot corrosion salt combinations at 1000 °C. In the case of YSZ top coat on TBCs, the reaction between Na_2SO_4, V_2O_5, and Y_2O_3 salts generates YVO_4 crystals, and these structures cause the transformation of tetragonal ZrO_2 to monoclinic ZrO_2. This situation occurs under operating conditions that lead to TBC failure. Hot corrosion behavior and the related failure mechanisms of TBC systems were investigated and discussed using scanning electron microscope (SEM), energy dispersive spectroscopy (EDS) analysis, and X-ray diffractometer (XRD).

Keywords: thermal barrier coatings (TBCs); electron beam physical vapor deposition (EB-PVD); cold gas dynamic spray (CGDS); hot corrosion; molten salts

1. Introduction

Thermal barrier coatings (TBCs) have been widely used to improve the efficiency of gas turbine engines in aviation industry [1–3]. Mostly, aircraft engines work at elevated temperatures, and these temperatures generally damage aircraft engine parts, such as turbines and blades. TBCs have been used in gas turbine engine components, like turbine blades and vanes, to protect them from aggressive environmental effects, increase operating temperatures, and provide thermal insulation at higher operating temperatures [4,5].

Yttria-stabilized zirconia (YSZ) is widely used as a traditional TBC ceramic top coat material due to its high melting point, low thermal conductivity, and high-temperature stability. MCrAlY alloy is also used as a typical TBC bond coat material due to its excellent high-temperature strength [1,2,4]. When TBCs are exposed to oxidation with high temperature, a thermally grown oxide (TGO) layer is formed between the top coat and bond coat [6]. The ceramic top coat has a substantially low thermal conductivity and partially low thermal expansion. The metallic bond coat is deposited between the metallic substrate and ceramic top coat for improving adherence of the ceramic top coating to the substrate alloy [7].

Bond coats are an important part of TBCs, and they can be produced by a variety of methods such as atmospheric plasma spray (APS), vacuum plasma spray (VPS), high-velocity oxy-fuel (HVOF), or cold gas dynamic spray (CGDS) [8]. CGDS is a thermal spraying technique by which the powder coating materials are deposited on the surface at high speeds at considerably lower process temperatures. This high-speed process is carried out by heating and compressing gases such as air, nitrogen, and helium. The CGDS method has many advantages over other thermal spray coating methods. The products produced by this method do not cause undesirable phase structures and oxidation. There is no change in the structure of the base material. It is also possible to obtain dense and high-hardness coatings. CGDS coatings can be produced at very high speeds. The most important advantage of this method is that CGDS method is applied at lower temperatures than other thermal spray coating processes, such as HVOF, APS, and D-gun [1,8]. CGDS technique may appear to be an alternative process to obtain superior coating properties for bond coat production. CGDS coatings, commonly known as "cold spray" coatings, exhibit remarkably high density, good corrosion resistance, and high hardness with high reliability, due to their cold-worked microstructure [9]. Cold-sprayed bond coats show improved adhesion to the substrate and provide a suitable surface roughness [10].

Ceramic top coat production is the most important part of TBC production. Generally, two methods are available for this purpose, namely, APS and EB-PVD. EB-PVD coatings are deposited in a columnar structure as opposed to APS coatings which grow in a laminar manner. EB-PVD has a more complex mechanism than plasma spraying, in terms of the equipment used. EB-PVD methods need preheating for the substrate while for plasma spray methods, it is not necessary to heat the substrate and the coating rate is faster than EB-PVD. APS is a low-cost, easy, and rapid production technique compared to EB-PVD. In addition, EB-PVD TBCs have good aerodynamic properties due to their surface roughness, which is better than APS [11].

TBCs undergo a variety of failure mechanisms, such as hot corrosion, oxidation, erosion, and thermal shock. Thermal shock occurs with sudden temperature alterations. Oxidation occurs as a result of chemical reactions of metallic components at high temperatures. Hot corrosion is the most important failure mechanism in TBC. Hot corrosion takes place due to the adverse effects of molten salts at high temperature [12]. In a working gas turbine, sulfur which comes from the low-quality fuel reacts with sodium chloride and sodium sulfate forms at high temperatures. After that, sodium sulfate covers the hot-section components. This phenomenon is commonly referred to as "hot corrosion". Turbine blade alloys that suffer hot corrosion attack have been widely found to show both oxidation and sulfidation [10]. This occurs at high temperature and in corrosive environments. However, this mechanism has not been exactly declared due to the use of low-quality fuels. Some low-quality fuels have impurities, such as Na_2SO_4 and V_2O_5 salts, covering the component surface at the high temperature. These salts reacting with the top coat cause phase transformation [1]. As a result, this transformation causes volume expansion and, accordingly, failure occurs depending on the TBC system [12].

In this study, the CGDS method, widely used as a novel technology in recent years, was used to deposit CoNiCrAlY-based metallic bond coats onto nickel-based superalloy substrate Inconel 718. EB-PVD method was used for deposition of ceramic top coatings with YSZ content, which are used for thermal insulation purposes. The primary aim of this study is the deposition of CoNiCrAlY-based metallic bond coats having a dense structure with low oxide and porosity content. Another aim is evaluation of the failure mechanisms and performances occurring under high temperature and corrosion conditions in EB-PVD TBC systems.

2. Experimental

2.1. Preparation of Substrate Material, and Bond and Top Coatings

Inconel 718, a nickel-based superalloy, was used as the substrate material. The approximate dimensions of 1 inch diameter and 4 mm thickness were obtained by cutting with wire erosion process. Sample surfaces were prepared by sandblasting for the coating process. The deposition of CoNiCrAlY metal powders was accomplished with CGDS technique. Approximately 100 μm coating thickness was achieved on nickel-based superalloy substrate material. Following the production of bond coat, the deposition of top coat was accomplished with nearly 300 μm thickness by EB-PVD technique using YSZ ceramic-based ingots. In the EB-PVD unit for deposition of top coating, YSZ ingot material of 68.5 mm diameter and 45–55 mm height was used.

2.2. Hot Corrosion Tests

Salt mixtures of Na_2SO_4 (99% purity) and V_2O_5 (98% purity) were scattered onto the surfaces of produced TBC samples. The corrosive salts were spread over the surface of specimens at a concentration of 10 mg/cm^2, keeping 1.5 mm from the edge to avoid edge effects. The samples were subjected to hot corrosion tests at 1000 °C (1273 K) temperature, with periodical cycling between furnace and room temperature for 4 h periods. These cycles were continued until about 40% deterioration was observed on the samples. After each 4 h period of testing at 1000 °C, the samples were allowed to cool down inside the furnace, and the coatings were then inspected for possible crack initiation. The samples were then recoated with the Na_2SO_4 + V_2O_5 salt mixture.

2.3. Characterization

Microstructures of TBCs were evaluated using a scanning electron microscope (SEM) (SEM, Tescan, MAIA3 XMU, Brno, Czech Republic) equipped with latest technology energy dispersive spectroscopy (EDS-SEM, Tescan, MAIA3 XMU). The phases of the TBCs were analyzed using X-ray diffraction (XRD) (Rigaku Dmax 2200 PC, Cu Kα radiation, Tokyo, Japan). Surface roughness values were measured using the surface roughness tester (SJ-310, Mitutoyo, Kanagawa, Japan). Hardness values of the coatings were measured using Vickers indentation using a Struers Duramin microhardness tester (Copenhagen, Denmark) with 100 g load for 15 s periods. Corrosive product formation and degradation mechanisms were examined using surface and cross-sectional microstructural analysis. The important results of this study are presented in this paper.

3. Results and Discussion

3.1. Characterization of as-Sprayed Coatings

Substrate material, metallic bond, and ceramic top coat's surface roughness were measured according to DIN EN ISO 3274 standard [13], and are shown in Table 1. Metallic bond coats, which were produced by CGDS methods, had dense, little porosity, and uniform structure [1,6,11].

Table 1. Surface roughness values of substrate, and bond and top coat.

Surface Roughness	R_a (μm)
Inconel-718 substrate	5.25
CGDS bond coat	7.20
EB-PVD top coat	7.92

The bond coat cross-section of the as-deposited EB-PVD TBCs with cold-sprayed CoNiCrAlY bond coat is demonstrated in Figure 1. As observed in the image, the coating exhibits a columnar structure, which is a typical feature of EB-PVD coating [1–5,14–17]. The TBC structure produced with EB-PVD method has a lower crack content. As-deposited TBC surface image is shown in Figure 2. The top surface of the coating exhibits a pyramidal appearance and intercolumnar gaps can be observed.

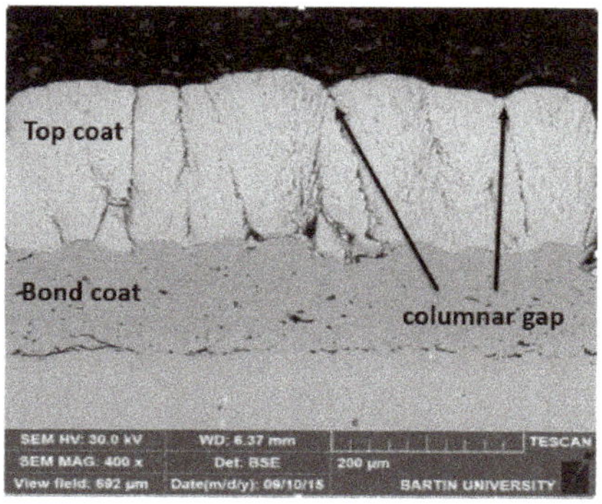

Figure 1. As-deposited cross-sectional microstructure of yttria-stabilized zirconia (YSZ) thermal barrier coating (TBC) system.

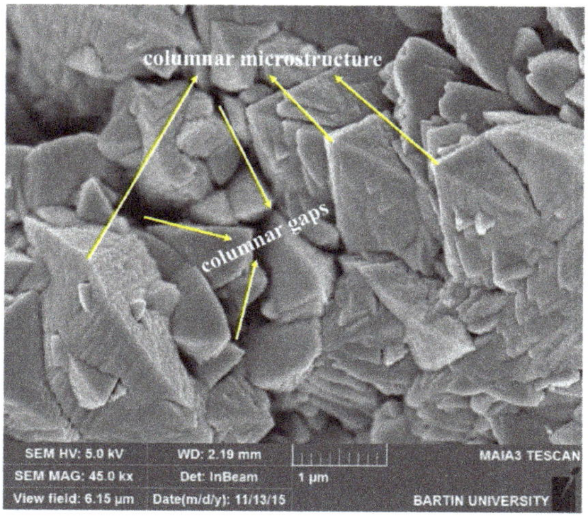

Figure 2. As-deposited columnar surface image micrograph of microstructure of YSZ TBC.

After hot corrosion tests, a uniform and dense coating structure provided by EB-PVD technique retards the formation of hot corrosion products [1,5,8,12]. The substrate material, metallic bond, and ceramic top coat surface hardness values are given in Table 2.

Table 2. Hardness values of the thermal barrier coating component.

Material	Hardness Values (H_v)
Inconel-718	203.9
CoNiCrAlY Bond Coat	407.9
YSZ Top Coat	713.8

3.2. *Effect of Hot Corrosion on TBC Samples*

Effects of hot corrosion occurred due to molten salts leaking from the microcracks and porosities in the top layer. Elements, such as sodium and vanadium, lead to a change in the phase structure of YSZ top coating layer in service conditions. If Na, S, P, V, etc. elements exist in structure due to fuel pollution, they can react with the Y_2O_3 phase. Tetragonal–monoclinic phase transformation in stabilized ZrO_2 happens in the absence of Y_2O_3 phase. This transformation causes both structural integrity deterioration and volumetric change [16,17]. Reaction of melted sulfate and vanadium salts is needed to completely understand the damage mechanism of hot corrosion on YSZ [11].

The process of hot corrosion has been emphasized in several earlier studies [1,6,11,14,15]. The hot corrosion action can be expressed by the following reactions [1,6,14,18,19].

$$V_2O_5\ (S) + Na_2SO_4\ (s) \rightarrow 2NaVO_3\ (S) + SO_3\ (g) \tag{1}$$

$$t - ZrO_2\ (Y_2O_3)\ (s) + 2NaVO_3\ (S) \rightarrow m - ZrO_2\ (s) + 2YVO_4\ (S) + Na_2O \tag{2}$$

The below reaction can also directly occur [1,6,14,18].

$$ZrO_2\ (Y_2O_3) + V_2O_5 \rightarrow ZrO_2\ (\text{monoclinic}) + 2YVO_4 \tag{3}$$

During the hot corrosion on TBCs, yttria slowly runs out. After depletion of yttria, the transformation from tetragonal zirconia to monoclinic zirconia during the cooling step of thermal cycling occurs by 3%–5% volume expansion, leading to cracking and spallation of TBCs [1,4,6,8,20].

A cross-section SEM image of EB-PVD TBC with cold-sprayed CoNiCrAlY bond coat after 20 h hot corrosion test is shown in Figure 3. No evidence of reaction was found between YSZ and Na_2SO_4 in the XRD results. However, Na_2SO_4 shows a faster effect of chemical reaction in hot corrosion mechanism [1,10,16,21]. Moreover, the presence of $NaVO_3$ increases the mobility of Y^{3+}. Thus, exhaustion of yttria and growth of YVO_4 crystal is enhanced [7]. It is accepted that YSZ usually stabilizes in tetragonal phase due to the presence of Y_2O_3. For this reason, in the case of a decrease in Y_2O_3, tetragonal ZrO_2 phase turns into monoclinic ZrO_2 phase [21–23]. At the interface between YSZ ceramic top coating and CoNiCrAlY metallic bond coating, TGO layer and surface cracks were formed. In addition, the EB-PVD TBCs exposed to hot corrosion salts exhibited spallation of the top coat near the bond coat/top coat interface. Monoclinic ZrO_2 and YVO_4 crystals caused wide crack formation in the YSZ ceramic top coating.

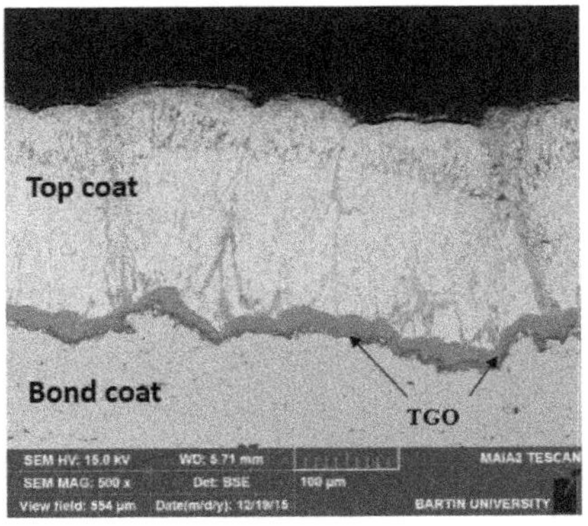

Figure 3. Cross-sectional microstructure of YSZ TBC after the hot corrosion test.

Following the hot corrosion test, closure of the porous structure was observed among the columns. This situation is attributed to the sintering effect and decreasing of spaces among the columns. When the undesirable effect of sintering appears as a damaging factor in TBC systems, decreased elongation tolerance at high temperatures due to increasing mechanical properties of the coating and consequent spallation damage occur [24–29]. After the 20 h hot corrosion test, the hot corrosion tests were stopped due to the separation of coating from the coating surface.

EB-PVD TBC surface image and EDS analysis after 20 h hot corrosion tests are respectively given in Figures 4 and 5. YVO_4 rod crystals in YSZ coating were revealed after hot corrosion. After the hot corrosion test, structures of rod-shaped crystals according to EDS and XRD analyses were identified with the formation of YVO_4 phases [6,21]. It was revealed that YSZ ceramic top coating consisted of a porous surface and a few rod crystals. The rod-shaped structures generally form as columnar boundaries on TBC samples. Y_2O_3 reacted with V_2O_5 and after this reaction, YVO_4 rod-shaped structures occurred on TBC samples [6].

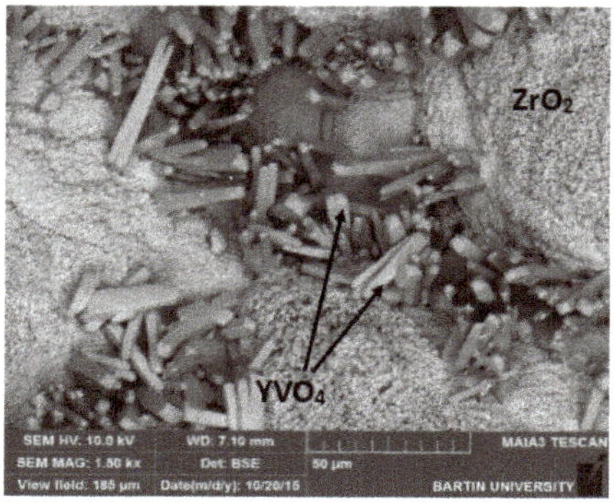

Figure 4. Surface microstructure of TBC after the hot corrosion test.

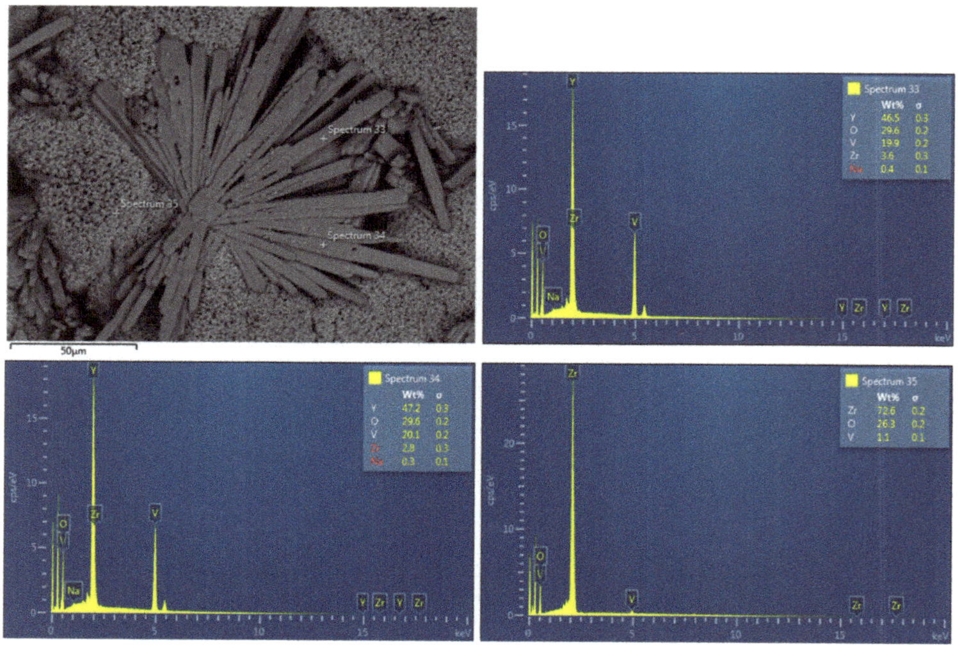

Figure 5. Surface microstructure with EDS analysis of YSZ TBC after the hot corrosion test.

In particular, due to the fact that columnar structure is denser and has less porosity, penetration of hot corrosion molten salts hardly occurred on the columnar microstructured TBCs. When the molten salts can easily penetrate from the top coat, a hot corrosion mechanism quickly occurs. Hot corrosion molten salt products, such as vanadium, oxygen, and yttria, can be seen in the cross-sectional image and surface image after hot corrosion tests. Hot corrosion products can be seen considering the surface image of samples in EDS analysis. Moreover, hot corrosion products can be seen in elemental mapping analysis (Figure 6) after hot corrosion test. According to that figure, oxygen coincides with vanadium and yttrium. It can be understood from this analysis that rod-shaped YVO_4 crystals were detected as one of the hot corrosion products. In addition to the EDS and elemental mapping analyses, the XRD phase analysis in Figure 7 demonstrates that monoclinic ZrO_2 phase is also another hot corrosion product besides the YVO_4 phase. Before the hot corrosion test, the top coat has only tetragonal ZrO_2 phase. After the 20 h hot corrosion test, top coat spallation occurs as a result of phase transformation.

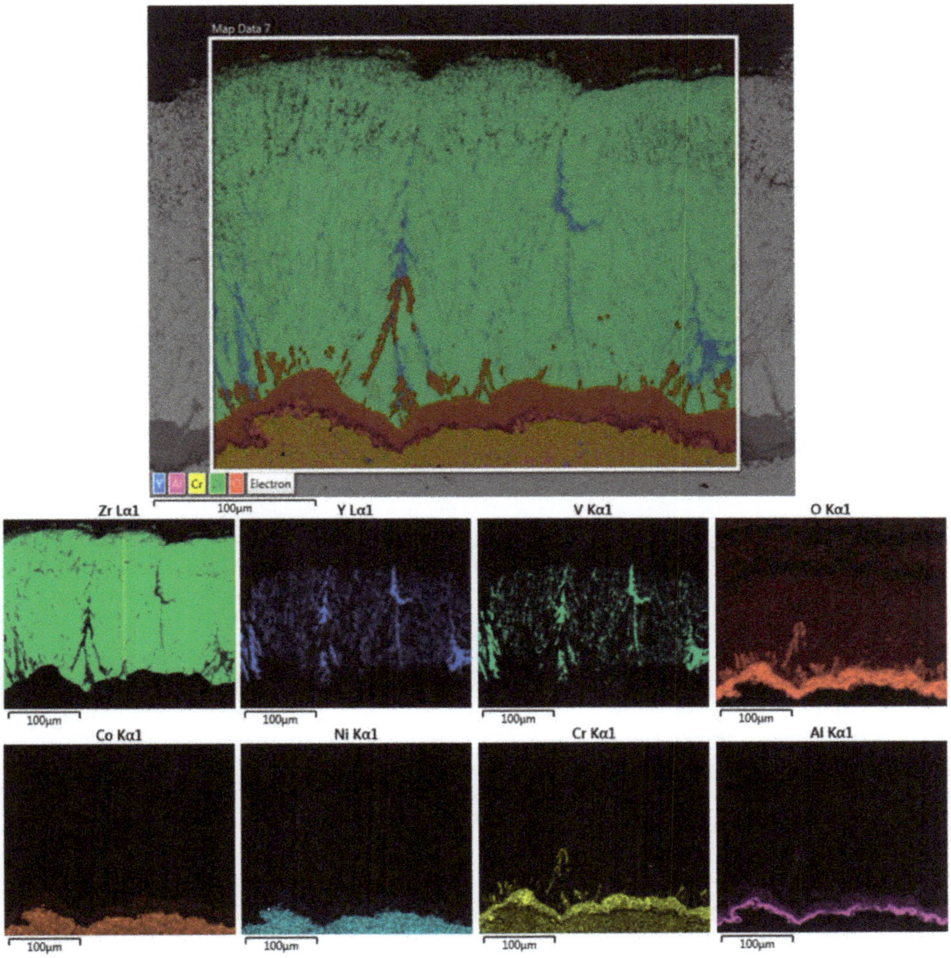

Figure 6. Cross-sectional microstructure with elemental mapping analysis belonging to TBC system after the hot corrosion test.

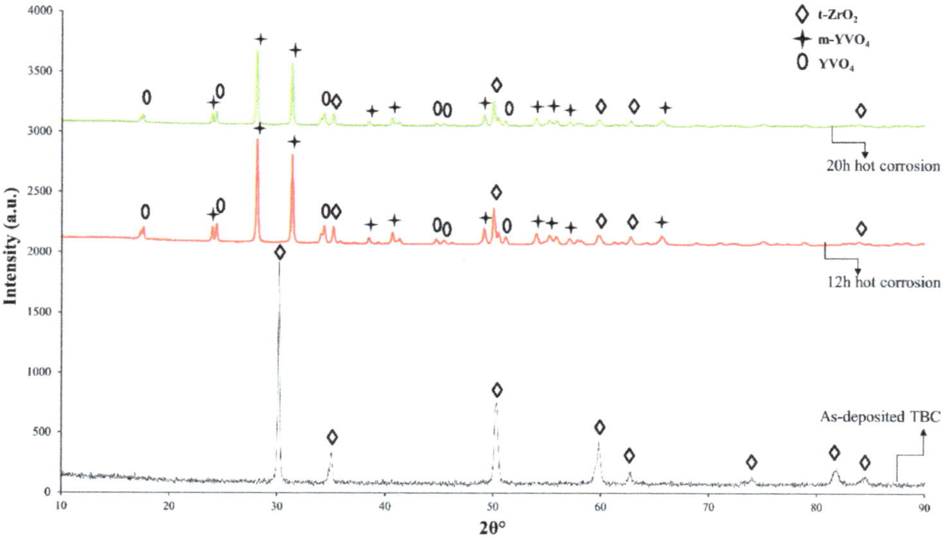

Figure 7. XRD analysis of TBC before and after hot corrosion tests.

During 22, 42, and 100 h hot corrosion tests in Na_2SO_4 + V_2O_5 salt mixture environment, formation of YVO_4 crystals were observed as hot corrosion products [5,6,16,23]. In similar studies conducted with YSZ top coating in Na_2SO_4 + V_2O_5 salt mixture environment as molten salt, phase transformation-related spallations were observed on top coatings of TBC samples, in spite of varying number and duration of cycles [1,6,16,21]. Due to the melting point of Na_2SO_4 + V_2O_5 salt mixture, about 900 °C, and that our test temperature is above this temperature, this salt mixture become molten during the test [18]. Hui et al. stated in their research [7] that after SEM, XRD, and EDS analyses, YVO_4 formations with rod-shaped crystal structure were observed as a result of isothermal hot corrosion tests with 6, 12, 24, 48, and 96 h time periods that were conducted at 1100 °C in a Na_2SO_4–30 mol % $NaVO_3$ environment. YVO_4 phases were only observed in 48 and 96 h periods, while monoclinic ZrO_2 was observed in all hot corrosion test periods according to XRD analysis. However, YVO_4 phases were obtained throughout each cycle in our present study. This may be related to the presence of V_2O_5 in our hot corrosion salts.

Liu et al. [30] observed a volumetric increase in thermal barrier coating samples comprising of Sc_2O_3–Y_2O_3–ZrO_2 (ScYSZ) subjected to hot corrosion test at 1000 °C in a Na_2SO_4 + V_2O_5 environment, which was attributed to the phase transformation of zirconium dioxide, from tetragonal to monoclinic phase [1,7,11,16,19,21]. This phase transformation caused crack formation in the coating. After the hot corrosion tests, it was determined that ScYSZ had a longer lifetime compared to YSZ. It is stated that the presence of less yttria in ScYSZ influences this situation. After corrosion tests of ScYSZ, phase transformation occurred in the coating structure, but no YVO_4 phase formation was observed. The EB-PVD TBC system provides a high elongation tolerance under thermal loads. This ensures that the thermal cycle lifetime of EB-PVD TBCs is longer than APS. However, the molten corrosion salts leaking from the columnar openings progress more rapidly. For this reason, while expansion columns are closing, corrosion products that react with the surfaces found here have an accelerating effect on the damage of the coating. In solutions of precursor plasma spray (SPSS), there are similar columnar openings. Ajay et al. [18] stated that in the case of the TBC samples produced with SPSS and APS methods, and tested in 90% Na_2SO_4 + 5% V_2O_5 + 5% NaCl environment and 50% Na_2SO_4 + 50% V_2O_5 environment at 900 °C, the sample produced with APS method exhibited a higher cycle life compared to the one produced with another method in both environments [3,14,21]. In other research [1,6,16,21,26] at higher temperatures such as 1050 and 1100 °C, and lower temperatures such as 950 °C, formations of rod-shaped YVO_4 were also observed in YSZ top coatings. However, in top coating materials other than YSZ, such as $Gd_2Zr_2O_7$ and Ta_2O_5, failures were observed in the form of spallations rather than rod-shaped formations like YVO_4 [1,6,16,21]. When Wang et al. [31] investigated $La_2Ce_2O_7$/YSZ thermal barrier double layer coatings, hot corrosion molten salts penetrated into the microcolumnar gap and reacted with La element. After the reaction of molten salt with La element, the top coating microstructure of $La_2Ce_2O_7$ deteriorated because of the formation of $CeVO_4$ and $LaVO_4$. In comparison with La2Ce2O7, YSZ was not affected by the chemical reaction of molten salts. $La_2Ce_2O_7$ stopped the progression of corrosive salts to YSZ [31]. Xu and others [32] used rare earth zirconates $La_2Zr_2O_7$ and $La_2(Zr_{0.7}Ce_{0.3})_2O_7$ to produce TBCs with EB-PVD method. TBC samples were subjected to hot corrosion test by holding them for 100 h in molten salt solution. In the XRD analyses performed after the hot corrosion tests, monoclinic zirconia and YVO_4 phases were found to be similar to our study. In a study conducted by Xu et al. [32], La_2O_3 served as a stabilizer in the coating. After the hot corrosion test, La_2O_3 within the structure was separated from the structure due to reaction with $NaVO_3$. Y_2O_3 reacted with $NaVO_3$ formed by molten salts and formed YVO_4 structure. This reaction led to tetragonal to the monoclinic phase transformation of the top coat. These reactions caused the damage in the top coat. Similar damage can be seen in this study [32].

Gavendova et al. [33] produced TBC coatings with CGDS and HVOF techniques, and reported that the interdendritic structure of the CGDS bond coats included more porosity as compared to those produced with HVOF method. Reportedly, the phase and chemical composition of HVOF bond coat was better than that of the CGDS bond coat. As opposed to Gavendova et al. [33], Knanna and Rathod [34] reported that CGDS coatings exhibited better wear and friction behavior than HVOF coatings. Owing to the low process temperatures of CGDS method, substrate materials do not undergo unfavorable structural changes during the deposition process. However, in their study on HVOF and CGDS, Khanna and Rathod showed that the coatings produced by the CGDS method showed better wear and friction behavior than the coatings produced by the HVOF method. As a result, the strength of the overall TBC system can be maintained by use of this method.

4. Conclusions

In the present work, the hot corrosion behavior of a new generation of TBC system, including a CoNiCrAlY bond coat and an 8YSZ top coat, was studied, and the following conclusions are drawn:

- The structure of metallic bond coat produced with CGDS technique was found to have a denser structure with lower porosity and oxide content due to the production characteristics of the process.
- Along with the hot corrosion effect, the formation of TGO was also observed on the specimens, since the furnace environment in which the hot corrosion tests were conducted was open to the atmosphere.
- The EDS analyses indicate that formations of rod crystal structures observed in the microstructural images obtained from the top surface of TBC systems belong to YVO_4 crystals. Chemical reactions of $NaVO_3$, forming as a result of the reaction of Y_2O_3, Na_2SO_4, and V_2O_5 salts penetrating into YSZ top coating, also contributed to the formation of YVO_4 crystals.
- The molten salts that leaked from the top coating caused deterioration of the structure with the columnar openings provided by EB-PVD, which played a role in accelerating damage. At the end of the hot corrosion tests, it was determined that the columnar structure was destroyed.
- Along with the YVO_4 crystals in rod form, the reasons underlying the failure of coatings in TBC systems were found to be volumetric changes and transformations at the rate of 3%–5% during the cooling process of ZrO_2, which has been transformed into monoclinic phase structure from tetragonal phase structure.

Author Contributions: A.C.K., Y.O., K.M.D., H.A. and D.O. designed the experiments; A.C.K. and Y.O. performed the experiments; A.C.K., Y.O., and K.M.D. analyzed the data; A.C.K., Y.O. and K.M.D. wrote, reviewed and edited the paper.

Funding: This research was funded by The Scientific and Technological Research Council of Turkey, TUBITAK, 113R049.

Acknowledgments: The authors thank H.E. (from Bartin University) for his support on characterization of samples.

Conflicts of Interest: The authors declare no conflict of interest.

References

1. Ozgurluk, Y.; Doleker, K.M.; Karaoglanli, A.C. Hot corrosion behavior of YSZ, $Gd_2Zr_2O_7$ and YSZ/$Gd_2Zr_2O_7$ thermal barrier coatings exposed to molten sulfate and vanadate salt. *Appl. Surf. Sci.* **2018**, *438*, 96–113. [CrossRef]
2. Sreedhar, G.; Raja, V.S. Hot corrosion of YSZ/Al_2O_3 dispersed NiCrAlY plasma-sprayed coatings in Na_2SO_4–10 wt.% NaCl melt. *Corros. Sci.* **2010**, *52*, 2592–2602. [CrossRef]
3. Eliaz, N.; Shemesh, G.; Latanision, R.M. Hot corrosion in gas turbine components. *Eng. Fail. Anal.* **2002**, *9*, 31–43. [CrossRef]
4. Clarke, D.R.; Levi, C.G. Materials design for the next generation thermal barrier coatings. *Annu. Rev. Mater. Res.* **2003**, *33*, 383–417. [CrossRef]
5. Karaoglanli, A.C.; Doleker, K.M.; Demirel, B.; Türk, A.; Varol, R. Effect of shot peening on the oxidation behavior of thermal barrier coatings. *Appl. Surf. Sci.* **2015**, *354*, 314–322. [CrossRef]
6. Habibi, M.H.; Wang, L.; Guo, S.M. Evolution of hot corrosion resistance of YSZ, $Gd_2Zr_2O_7$, and $Gd_2Zr_2O_7$ + YSZ composite thermal barrier coatings in Na_2SO_4 + V_2O_5 at 1050 °C. *J. Eur. Ceram. Soc.* **2012**, *32*, 1635–1642. [CrossRef]
7. Hui, Y.; Zhao, S.; Xu, J.; Zou, B.; Wang, Y.; Cai, X.; Zhu, L.; Cao, X. High-temperature corrosion behavior of zirconia ceramic in molten Na_2SO_4 + $NaVO_3$ salt mixture. *Ceram. Int.* **2016**, *42*, 341–350. [CrossRef]
8. Moy, C.K.S.; Cairney, J.; Ranzi, G.; Jahedi, M.; Ringer, S.P. Investigating the microstructure and composition of cold gas-dynamic spray (CGDS) Ti powder deposited on Al 6063 substrate. *Surf. Coat. Technol.* **2010**, *204*, 3739–3749. [CrossRef]
9. Kuzmichew, A.; Tsybulski, L. Evaporators with induction heating and their applications. In *Advances in Induction and Microwave Heating of Mineral and Organic Materials*; Grundas, S., Ed.; IntechOpen: London, UK, 2011; pp. 269–302.
10. Burlacov, I.; Jirkovský, J.; Kavan, L.; Ballhorn, R.; Heimann, R.B. Cold gas dynamic spraying (CGDS) of TiO_2 (anatase) powders onto poly(sulfone) substrates: Microstructural characterisation and photocatalytic efficiency. *J. Photochem. Photobiol. A Chem.* **2007**, *187*, 285–292. [CrossRef]
11. Wahl, G.; Metz, C.; Samoilenkov, S. Thermal barrier coatings. *J. Phys. IV France* **2001**, *11*, Pr3-835–Pr3-846. [CrossRef]
12. Jiang, S.M.; Peng, X.; Bao, Z.B.; Liu, S.C.; Wang, Q.M.; Gong, J.; Sun, C. Preparation and hot corrosion behaviour of a MCrAlY + AlSiY composite coating. *Corros. Sci.* **2008**, *50*, 3213–3220. [CrossRef]
13. *ISO 3274:1996 Geometrical Product Specifications (GPS)–Surface Texture: Profile Method–Nominal Characteristics of Contact (Stylus) Instruments*; International Organization for Standardization: Geneva, Switzerland, 1996.
14. Afrasiabi, A.; Saremi, M.; Kobayashi, A. A comparative study on hot corrosion resistance of three types of thermal barrier coatings: YSZ, YSZ + Al_2O_3 and YSZ/Al_2O_3. *Mater. Sci. Eng. A* **2008**, *478*, 264–269. [CrossRef]
15. Habibi, M.H.; Wang, L.; Liang, J.; Guo, S.M. An investigation on hot corrosion behavior of YSZ-Ta_2O_5 in Na_2SO_4 + V_2O_5 salt at 1100 °C. *Corros. Sci.* **2013**, *75*, 409–414. [CrossRef]
16. Daroonparvar, M.; Yajid, M.A.M.; Yusof, N.M.; Bakhsheshi-Rad, H.R.; Hamzah, E.; Nazoktabar, M. Investigation of three steps of hot corrosion process in Y_2O_3 stabilized ZrO_2 coatings including nano zones. *J. Rare Earths* **2014**, *32*, 989–1002. [CrossRef]
17. Mensch, A.; Thole, K.A.; Craven, B.A. Conjugate heat transfer measurements and predictions of a blade endwall with a thermal barrier coating. *J. Turbomach.* **2014**, *136*, 121003. [CrossRef]

18. Ajay, A.; Raja, V.S.; Sivakumar, G.; Joshi, S.V. Hot corrosion behavior of solution precursor and atmospheric plasma sprayed thermal barrier coatings. *Corros. Sci.* **2015**, *98*, 271–279. [CrossRef]
19. Mifune, N.; Harada, Y.; Doi, T.; Yamasaki, R. Hot-corrosion behavior of graded thermal barrier coatings formed by plasma-spraying process. *J. Therm. Spray Technol.* **2004**, *13*, 561–569. [CrossRef]
20. Mobarra, R.; Jafari, A.H.; Karaminezhaad, M. Hot corrosion behavior of MCrAlY coatings on IN738LC. *Surf. Coat. Technol.* **2006**, *201*, 2202–2207. [CrossRef]
21. Levi, C.G.; Vidal-Sétif, M.H.; Johnson, C.A.; Hutchinson, J.W. Environmental degradation of thermal-barrier coatings by molten deposits. *MRS Bull.* **2012**, *37*, 932–941. [CrossRef]
22. Tsai, P.C.; Lee, J.H.; Hsu, C.S. Hot corrosion behavior of laser-glazed plasma-sprayed yttria-stabilized zirconia thermal barrier coatings in the presence of V_2O_5. *Surf. Coat. Technol.* **2007**, *201*, 5143–5147. [CrossRef]
23. Li, S.; Liu, Z.G.; Ouyang, J.H. Growth of $YbVO_4$ crystals evolved from hot corrosion reactions of $Yb_2Zr_2O_7$ against V_2O_5 and $Na_2SO_4 + V_2O_5$. *Appl. Surf. Sci.* **2013**, *276*, 653–659. [CrossRef]
24. Yugeswaran, S.; Kobayashi, A.; Ananthapadmanabhan, P.V. Hot corrosion behaviors of gas tunnel type plasma sprayed $La_2Zr_2O_7$ thermal barrier coatings. *J. Eur. Ceram. Soc.* **2012**, *32*, 823–834. [CrossRef]
25. Feuerstein, A.; Knapp, J.; Taylor, T.; Ashary, A.; Bolcavage, A.; Hitchman, N. Technical and economical aspects of current thermal barrier coating systems for gas turbine engines by thermal spray and EBPVD: A review. *J. Therm. Spray Technol.* **2008**, *17*, 199–213. [CrossRef]
26. Ghasemi, R.; Shoja-Razavi, R.; Mozafarinia, R.; Jamali, H. The influence of laser treatment on thermal shock resistance of plasma-sprayed nanostructured yttria stabilized zirconia thermal barrier coatings. *Ceram. Int.* **2014**, *40*, 347–355. [CrossRef]
27. Saremi, M.; Valefi, Z.; Abaeian, N. Hot corrosion, high temperature oxidation and thermal shock behavior of nanoagglomerated YSZ–alumina composite coatings produced by plasma spray method. *Surf. Coat. Technol.* **2013**, *221*, 133–141. [CrossRef]
28. Kotlan, J.; Ctibor, P.; Pala, Z.; Homola, P.; Nehasil, V. Improving dielectric properties of plasma sprayed calcium titanate ($CaTiO_3$) coatings by thermal annealing. *Ceram. Int.* **2014**, *40*, 13049–13055. [CrossRef]
29. Mohan, P. Environmental Degradation of Oxidation Resistant and Thermal Barrier Coatings for Fuel-Flexible Gas Turbine Applications. Ph.D. Thesis, University of Central Florida, Orlando, FL, USA, May 2010.
30. Liu, H.F.; Xiong, X.; Li, X.B.; Wang, Y.L. Hot corrosion behavior of Sc_2O_3–Y_2O_3–ZrO_2 thermal barrier coatings in presence of $Na_2SO_4 + V_2O_5$ molten salt. *Corros. Sci.* **2014**, *85*, 87–93. [CrossRef]
31. Wang, X.; Guo, L.; Peng, H.; Zheng, L.; Guo, H.; Gong, S. Hot-corrosion behavior of a $La_2Ce_2O_7$/YSZ thermal barrier coating exposed to $Na_2SO_4 + V_2O_5$ or V_2O_5 salt at 900 °C. *Ceram. Int.* **2015**, *41*, 6604–6609. [CrossRef]
32. Xu, Z.; He, L.; Mu, R.; He, S.; Huang, G.; Cao, X. Hot corrosion behavior of rare earth zirconates and yttria partially stabilized zirconia thermal barrier coatings. *Surf. Coat. Technol.* **2010**, *204*, 3652–3661. [CrossRef]
33. Gavendová, P.; Čížek, J.; Čupera, J.; Hasegawa, M.; Dlouhý, I. Microstructure modification of CGDS and HVOF sprayed CoNiCrAlY bond coat remelted by electron beam. *Procedia Mater. Sci.* **2016**, *12*, 89–94. [CrossRef]
34. Khanna, A.S.; Rathod, W.S. Development of CoNiCrAlY oxidation resistant hard coatings using high velocity oxy fuel and cold spray techniques. *Int. J. Refract. Met. Hard Mater.* **2015**, *49*, 374–382. [CrossRef]

coatings

Article

Deposition of $SiO_xC_yH_z$ Protective Coatings on Polymer Substrates in an Industrial-Scale PECVD Reactor

Žiga Gosar [1,2], Janez Kovač [3], Miran Mozetič [3,4], Gregor Primc [3,4], Alenka Vesel [3,4] and Rok Zaplotnik [3,4,*]

1 Elvez Ltd, Ulica Antona Tomšiča 35, 1294 Višnja Gora, Slovenia; ziga.gosar@elvez.si
2 Ecotechnologies, Jozef Stefan International Postgraduate School, Jamova cesta 39, 1000 Ljubljana, Slovenia
3 Department of Surface Engineering, Jozef Stefan Institute, Jamova cesta 39, 1000 Ljubljana, Slovenia; janez.kovac@ijs.si (J.K.); miran.mozetic@ijs.si (M.M.); gregor.primc@ijs.si (G.P.); alenka.vesel@ijs.si (A.V.)
4 Plasmadis Ltd, Teslova ulica 30, 1000 Ljubljana, Slovenia
* Correspondence: rok.zaplotnik@ijs.si

Received: 6 March 2019; Accepted: 1 April 2019; Published: 3 April 2019

Abstract: The deposition of protective coatings on aluminised polymer substrates by a plasma enhanced chemical vapour deposition PECVD technique in a plasma reactor with a volume of 5 m^3 was studied. HMDSO was used as a precursor. Plasma was sustained in a capacitively coupled radiofrequency (RF) discharge powered by an RF generator operating at 40 kHz and having an adjustable output power up to 8 kW. Gaseous plasma was characterised by residual gas mass spectrometry and optical emission spectroscopy. Polymer samples with an average roughness of approximately 5 nm were mounted into the plasma reactor and subjected to a protocol for activation, metallisation and deposition of the protective coating. After depositing the protective coating, the samples were characterised by secondary ion mass spectrometry (SIMS) and X-ray photoelectron spectroscopy (XPS). The combination of various techniques for plasma and coating characterisation provided insight into the complex gas-phase and surface reactions upon deposition of the protective coatings in the industrial-size plasma reactor.

Keywords: plasma-enhanced chemical vapour deposition (PECVD); hexamethyldisiloxane (HMDSO); industrial-size plasma reactor; capacitively coupled radiofrequency (RF) discharge; optical emission spectroscopy (OES); time-of-flight secondary ion mass spectrometry (ToF-SIMS); X-ray photoelectron spectroscopy (XPS)

---

1. Introduction

Plasma-enhanced chemical vapour deposition (PECVD) is a popular technique for the deposition of thin films, including protective coatings. Over 8000 scientific papers with a keyword PECVD have been published, including 400 papers in 2018. Although the technique was commercialised decades ago, it is still of high scientific importance despite its simple basic concept: a gaseous precursor is introduced into a processing chamber where it is weakly ionised and partially radicalised under plasma conditions, and the resultant reactive species condense on the substrate and thus cause growth of a solid film. Traditionally, the processing takes place in a vacuum chamber, but the current trend is applying atmospheric-pressure discharges [1] due to their obvious advantage: they are much cheaper than their vacuum counterparts.

For the deposition of Si-containing films, various precursors can be used [2–4], one of the most popular of which is hexamethyldisiloxane (HMDSO). It is used for the deposition of various coatings ranging from polymers to almost pure SiO_x films [5,6]. HMDSO is liquid at atmospheric pressure, but the vapour pressure is rather large (approximately 50 mbar at room temperature); therefore, it is suitable for introduction into a processing chamber via a leak valve or a gas flow controller. Non-equilibrium gaseous plasma is sustained in the processing chamber, and free electrons (whose temperature is on of the order of 10,000 K) cause radicalisation and ionisation of the precursor. Plasma is often sustained by high-frequency discharges of rather low power density. In a typical industrial reactor, the power density is on the order of kW/m^3. The reactive species diffuse inside the chamber and eventually reach the surface, where they stick at a certain probability. The substrates are often activated prior to deposition to improve the adhesion of the coating. Depending on plasma parameters, coatings of various properties are deposited.

One extreme is a coating resembling polydimethylsiloxane. Such a coating is obtained at a very low power density (to preserve the original composition of HMDSO) and with almost no other gases present. The other extreme is a thin film of silicon dioxide that grows when an oxygen-containing gas is present in the chamber. Water vapour is usually in chambers, and, occasionally, a small amount of oxygen is intentionally added to favour the formation of almost pure SiO_x. In between these two extremes, numerous types of coatings are obtainable depending on the processing parameters. The flux of radicals onto the surface (and thus the deposition rate) obviously increases with increasing pressure and power density. At elevated pressure (more than approximately 10 Pa), the radicals, however, start to agglomerate in the gas phase, so the coating becomes grainy, which is often regarded as detrimental in industrial systems. If the power density is elevated, the dissociation of the precursor is comprehensive; therefore, carbon atoms or even dimers may be incorporated into the SiO_x film, making it less transparent, which is also a major obstacle in such systems. Another obstacle is plasma uniformity—the dissociation and ionisation events are more intensive next to the electrodes, so the SiO_x film is deposited primarily onto the electrodes instead of on the substrates. The tuning of discharge and plasma parameters therefore represents a challenge that has been the focus of scientific study for decades. Only a recent literature survey is presented here.

Mitschker et al recently presented a good summary on the formation of protective coatings (see the references in paper [7]) and stressed the influence of oxygen admixture on the barrier properties of deposited films. He also reported beneficial results using HMDSN instead of HMDSO. Another rather complete paper on coatings prepared by PECVD using HMDSO was published recently by Prykril et al. [8]. They found an immeasurably low coefficient for oxygen permeation for a polymer foil coated with a SiO_2 layer. They also investigated the brittleness of the glassy coating and stressed that SiO_x must be deposited primarily on a low-swellability polymeric foil and that, due to its brittleness, it must immediately be laminated with another foil to prevent mechanical damage. M. Cavalcante dos Reis [9] reported interesting results on properties of coatings on dental materials. Various discharge parameters were used to obtain coatings with appropriate properties. This paper represents an interesting application in the niche characterised by a high value-added product. An interesting approach to the deposition of SiO_x coatings was adopted by Brochhagen et al. [10]. Instead of using a continuous supply of precursor, they pulsed small doses of HMDSO into the reaction chamber and used oxygen or argon plasma for transformation of the extremely thin polymer film created on the surface at each HMDSO pulse into a silica-like protective coating. This approach led to the formation of SiO_x films free from carbon or Si–OH groups with a very high (for industrial standards) deposition rate of approximately 1 nm/s. Scalability might be questionable since inductively coupled RF plasma in H-mode was applied. Pandiyaraj et al. [11] reported a DC plasma source for the rapid deposition (approximately 1 nm/s) of SiO_x films using a mixture of HMDSO and oxygen at a total pressure of 20 Pa. They found an interesting evolution of the film morphology as a function of discharge parameters. The initial stages in the nucleation and growth of SiO_x films on model substrates were elaborated by a team from Paderborn and Bochum [12]. Unlike most studies, they used oxygen as a major gas with a small admixture of HMDSO. Experiments were performed in a pulsed microwave plasma reactor at 25 Pa. The group also reported a numerical value of the O-atom flux onto the surface in order to explain the surface reactions in a rather quantitative manner. The key observation was a high density of defects, which was attributed to the incorporation of oxidised aliphatic fragments into the SiO_x network and/or the deterioration of the plasma-induced crosslinking of adsorbed HMDSO fragments.

Atmospheric pressure discharges have also been used for deposition of such coatings from HMDSO. Gas-phase and surface effects were studied both theoretically and experimentally. Among recent papers, D. Loffhagen et al. [13] applied a time-dependent, spatially one-dimensional fluid-Poisson model to analyse the impact of HMDSO admixtures on the discharge characteristics of DBDs in argon at atmospheric pressure. They considered a limited number of collision processes relevant for HMDSO but obtained reasonable consistency with available experimental data. They found that Penning ionisation due to collisions of HMDSO with metastable and resonance argon atoms becomes the main source of electron production already for HMDSO fractions ≥5 ppm. Some attempts have also been made to coat materials immersed into liquid HMDSO and apply atmospheric-pressure plasma next to the liquid surface. The most recent paper was published by Pavlinak et al. [14], who created plasma above liquid HMDSO and successfully coated Teflon tubes with a thin film resembling polydimethylsiloxane. Obviously, such treatments cannot be used for the deposition of SiO_x coatings.

Thin films containing Si, O, C and H are suitable for numerous applications. Traditionally, these films have been used as hydrophobic and flexible protective coatings on products exposed to harsh environments. A thin PDMS film is an effective coating to prevent corrosion of metals [15]. It is also a classic textile coating able to ensure the super-hydrophilic properties of textiles [16]. More recently, it has been used as a scaffold for tissue engineering and similar biomedical applications [17–19], including microfluidic systems [20,21]. A comprehensive review of recent applications was reported by Edouk et al. [22].

Only recent achievements have been cited above. The literature on the PECVD deposition of protective coatings is extensive, and various authors have reported results obtained in different experimental setups. Although the technique was commercialised decades ago, the scientific literature on the kinetics of deposition with HMDSO in industrial-scale reactors is very scarce. The present paper fills this niche since we report a thorough investigation of film deposition using PECVD with HMDSO as a precursor in a large-scale reactor that is used for the deposition of protective coatings on car lamps. Unlike most other authors, we performed experiments under conditions typical for industrial reactors. In such reactors, the residual atmosphere always plays an important role due to extensive desorption of gaseous molecules from the surfaces of numerous components mounted inside the system. Nitrogen, oxygen, carbon dioxide and argon (air constituents) are quickly pumped from the system by powerful pumps, but water molecules slowly desorb from the polymer components. The consequence is a rather substantial amount of water vapour in the system throughout the process; therefore, one should always take into account the effect of water vapour when studying processing parameters and explaining the interaction of gaseous plasma with solid materials. Another important difference between experimental systems and large reactors is the power density (discharge power per unit plasma volume). While it is not feasible to operate experimental reactors at powers below few 10 W per litre (few 10 kW/m^{-3}), the industrial reactors suitable for coating components with a thin film resembling polydimethylsiloxane operate at power densities below 1 kW/m^{-3}.

2. Experimental

The plasma reactor used in this study is intended for coating polymer components of rather complex shapes and typical dimensions on the order of 10 cm. The coating procedure is shown in Figure 1. The reactor was first pumped to a pressure of approximately 1 Pa to evacuate air. After pumping for approximately 4 min, argon was introduced upon continuous pumping so that a pressure of approximately 6 Pa was achieved. Plasma was sustained for approximately 3 min to ensure appropriate activation of the polymer components. Argon was then pumped away, and high-vacuum pumps were used to establish high-vacuum conditions. As the pressure dropped to approximately 0.02 Pa, aluminium was evaporated, so the polymer components were coated with a thin metal film.

As soon as evaporation was accomplished, HMDSO was introduced, and the plasma was ignited to deposit a thin film resembling polydimethylsiloxane (PDMS).

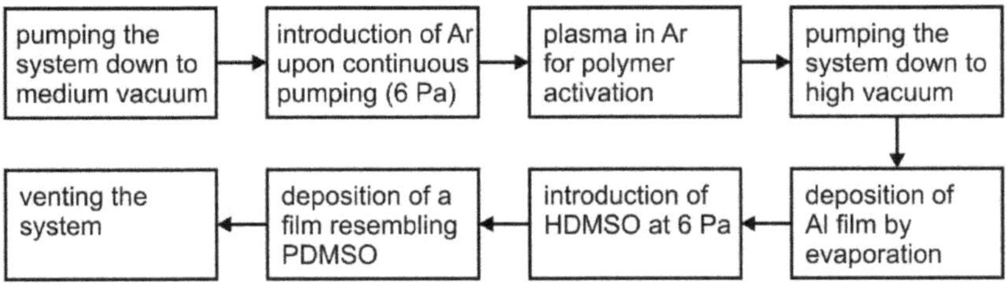

Figure 1. Schematic of the coating procedure.

The experimental system is shown in detail in Figure 2. The plasma reactor was a large chamber of a cylindrical shape made from steel. The diameter of the plasma reactor was 1.9 m, and the height was 1.8 m. The volume of the reactor was thus approximately 5 m^3. The reactor was pumped with a couple of diffusion pumps with a total nominal pumping speed of approximately 40 m^3/s. The fore-pumps were large one-stage oil rotary pumps with a nominal pumping speed of 1260 m^3/h. In between these pumps, there were a couple of roots blowers with a combined nominal pumping speed of 8800 m^3/h to ensure pre-compression of gas before entering the large rotary pump and thus enable an appropriately low pressure at the exhaust from the diffusion pumps. A cold trap was cooled down to a temperature of approximately 130 K and represents an extremely powerful pump for pumping water vapour from the plasma reactor. The nominal pumping speed of the cold trap was 140 m^3/s. HMDSO was introduced into the plasma reactor via a flow controller. The HMDSO container was kept at a constant temperature of approximately 310 K. There was also an argon flask connected to the plasma reactor via a flow controller. Plasma was ignited and sustained with an RF generator. The generator operated at a frequency of 40 kHz and an adjustable output power up to approximately 10 kW. The generator was connected to the RF electrode, which was placed asymmetrically in the plasma reactor. The dimensions of the electrode were 150 cm × 27 cm, so the electrode area was much smaller than the area of the grounded plasma reactor. The discharge coupling was almost purely capacitive. The asymmetry caused preferential power dissipation in front of the powered electrode. The electrode was subjected to ion bombardment, while the bombardment of the grounded housing was regarded marginal. Samples were mounted on fixtures and kept at floating potential during plasma treatment. Typical samples were polymeric components. There was a holder of small aluminium sheets that were heated under high-vacuum conditions by resistive heating to allow rapid deposition of a thin aluminium film onto the surface of the polymeric samples. Plasma was characterised by a residual atmosphere gas analyser (RGA, PrismaPlus QMG 220, Pfeiffer Vacuum, Asslar, Germany, sometimes referred to as mass spectrometry) and optical spectroscopy. The RGA was differentially pumped using a turbomolecular pump HiPace 700 (Pfeiffer Vacuum, Asslar, Germany) with a nominal pumping speed of 0.7 m^3/s backed with a small two-stage rotary pump with a nominal pumping speed of 5 m^3/h. There was a glass window on the plasma reactor, which was made from steel. An optical fibre was mounted onto the glass window and connected to an optical spectrometer AvaSpec-3648 (Avantes, Apeldoorn, The Netherlands) to allow the sampling of optical spectra from gaseous plasma in the wavelength range between approximately 200 and 1100 nm. The transmission of the glass window was negligible in both the UV and IR ranges.

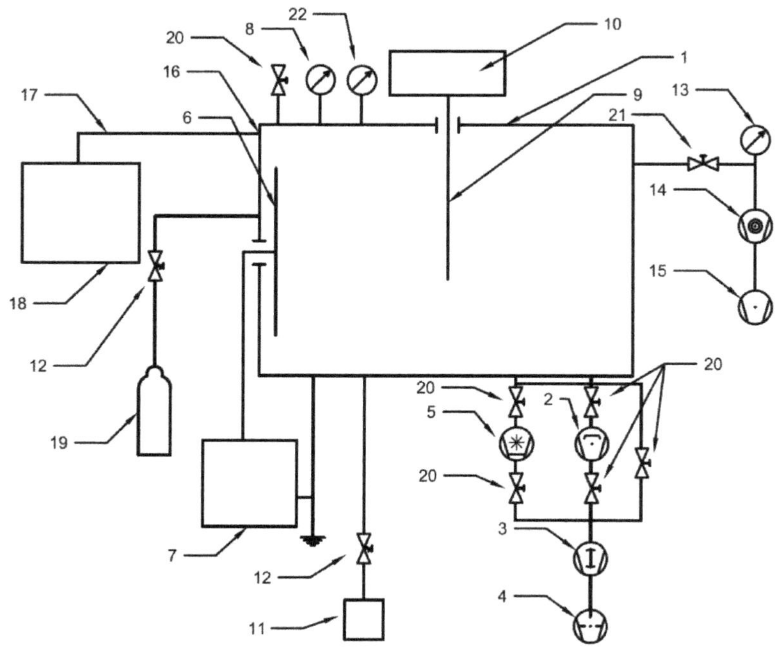

Figure 2. Schematic of the experimental setup: 1, plasma reactor; 2, diffusion pumps; 3, roots pumps; 4, large rotary pumps; 5, cold trap; 6, RF electrode; 7, RF generator; 8, Pirani gauge; 9, Al holder; 10, power supply for Al evaporation; 11, HMDSO container; 12, flow controller; 13, residual gas analyser; 14, turbomolecular pump; 15, small rotary pump; 16, glass window; 17, optical fibre; 18, optical spectrometer; 19, Argon container; 20, valves; 21, dosing valve; 22, Penning gauge.

Samples were commercial plastic components made from polycarbonate. The samples were made by injection moulding, and the surface of each sample was approximately 200 cm^2. Several samples were cut into small pieces with a surface area of approximately 1–2 cm^2 to enable characterisation. The samples were taken directly from the production line, so no pre-cleaning was performed. Some samples were first characterised by atomic force microscopy (AFM, Solver PRO, NT-MDT, Moscow, Russia) to determine their roughness. The average roughness (S_a), as determined by AFM, over an area of 20 µm × 20 µm was approximately 5 nm or, in terms of root-mean-square height, S_q = 6.5 nm. An AFM image is presented in Figure S1 in the Supplementary Materials. The samples were then mounted into the plasma reactor. The reactor was pumped to a reasonable pressure using the roots and large rotary pumps. Then, some argon was introduced into the reactor during continuous pumping, and plasma was sustained in the mixture of argon and residual atmosphere for a couple of minutes. This treatment was necessary for surface activation of the polymer samples prior to deposition of any coating. After plasma activation, the valve between the plasma reactor and the diffusion pumps was opened to reach high-vacuum conditions. The final pumping before depositing an aluminium coating was also performed with the cold trap. Once a suitable vacuum level was achieved, a thin film of aluminium was deposited in a short time to prevent significant contamination of the Al film due to the residual atmosphere. As soon as the Al coating was prepared, the cold trap and diffusion pumps were separated from the plasma reactor, HMDSO was introduced through the flow controller, and discharge was turned on to deposit the protective coating. During the introduction of the precursor, the reactor was pumped only with the roots and large rotary pumps. After depositing the protective coating, the pumps were separated, the reactor was vented, and the samples were taken for characterisation.

Samples with protective coatings were characterised by time-of-flight secondary ion mass spectrometry (SIMS) and X-ray photoelectron spectroscopy (XPS). In both cases, we used ion guns for removal of surface layers to obtain depth profiles. We used a ToF–SIMS 5 instrument (ION-TOF, Münster, Germany) equipped with a bismuth liquid metal ion gun with a kinetic energy of 30 keV. SIMS spectra were measured by scanning a Bi_3^+ cluster ion beam over an area of 100 × 100 μm^2. The beam current was 0.6 pA and the total measurement time to acquire the SIMS spectra was 30 s. The dose of primary ions during the measurements was in the static regime. The SIMS spectra were processed with the software SurfaceLab 6.3 (ION TOF).

Some samples were also characterised by XPS. We used a PHI-TFA XPS spectrometer produced by Physical Electronics Inc. (Chanhassen, MN, USA). The analysed area was 0.4 mm in diameter, and the analysed depth was approximately 3–5 nm. The sample surface was excited by X-ray radiation from a monochromatic Al source at a photon energy of 1486.6 eV. High-energy resolution spectra were acquired with an energy analyser operating at a resolution of approximately 0.6 eV and a pass energy of 29 eV. During data processing, the spectra acquired from the surface were aligned by setting the C 1*s* peak at 284.8 eV, characteristic of C–C bonds. Quantification of surface composition was performed on XPS peak intensities taking into account relative sensitivity factors provided by the instrument manufacturer.

3. Results and Discussion

The pressure in the plasma reactor presented schematically in Figure 2 during the entire procedure for deposition of protective coatings is shown in Figure 3. The reactor was first pumped with the large rotary pump. When a pressure of approximately 5000 Pa was reached, the roots pump was turned on, resulting in a knee in the curve at 1 min. The second knee was at approximately 50 Pa when cold trap was turned on, and the third knee was at approximately 5 Pa when valves to the diffusion pumps were opened. When the pressure dropped to approximately 1 Pa, argon was introduced, so the pressure increased to approximately 6 Pa. Gaseous plasma was created in the mixture of Ar and residual atmosphere to ensure appropriate surface activation. This treatment was accomplished in minute 6, as indicated in Figure 3. Then, the reactor was pumped with the cold trap and diffusion pumps for approximately 6 min. When the pressure inside the reactor dropped to a level of approximately 0.02 Pa, the Al heater was turned on (13 min). Once the Al coating was obtained, the cold trap and the diffusion pumps were separated, causing an almost instantaneous increase in pressure. HMDSO was then introduced, so the pressure increased to approximately 6 Pa. After that, the three cycles of the PECVD process were performed.

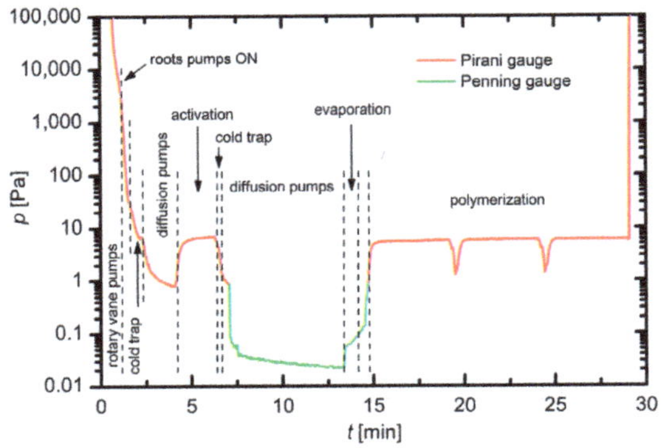

Figure 3. The vacuum levels in the plasma reactor at different stages.

During the upper procedure, mass spectra were measured continuously. Figure 4 presents the behaviour of different masses versus time in terms of the ion currents. The gaseous molecules were partially ionised in the residual gas analyser, and the positively charged ions entered the ion detector and caused ion currents. Comparison of Figures 3 and 4 reveals important observations that were crucial for deposition of the protective coatings. During initial pumping, there were essentially ion currents at only masses 18 and 16. These masses correspond to water and methane, respectively. The absence of currents at masses 28 and 32 indicated the lack of nitrogen and oxygen, respectively, in the vacuum chamber, which was explained by two facts: first, the plasma reactor was hermetically tight, and second, air was pumped much more efficiently than water vapour. The latter was due to extensive release of water vapour from the polymer samples. The intensity of water was considerably larger than that of methane, so the residual atmosphere in the plasma reactor practically consists of water vapour only.

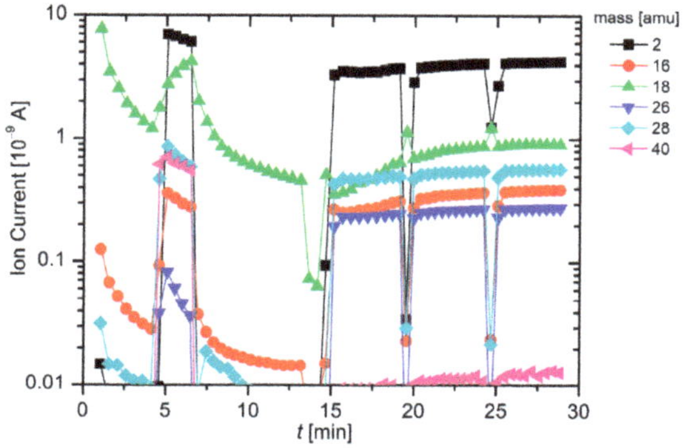

Figure 4. Mass spectrometry of gaseous molecules during all phases of the production process.

When argon was introduced into the reactor, there was an appearance of mass 40 in the spectrum presented in Figure 4. Simultaneously, as the discharge was turned on, one can observe the appearance of mass 2, which was attributed to hydrogen. The water vapour was partially dissociated under plasma conditions to H and OH radicals. H_2 molecules were probably formed by heterogeneous surface recombination of H atoms on the wall of the discharge chamber as well on the polymer samples. Extensive recombination also appeared on the path between plasma and the RGA—this connection was narrow, thus numerous collisions of gaseous species with the surface of the steel tube appeared. The recombination coefficients for polymeric materials were rather low [23], but for H atoms on stainless steel, the probability of surface recombination was close to 0.1 [24]; thus, very few H atoms created in plasma could enter the RGA. As a result, the H_2 signal prevails in Figure 4. Here, it is also worth mentioning that the sensitivity of the RGA for hydrogen was particularly large, thus no quantification of results in terms of partial pressure was possible.

During the activation of polymer substrates with argon plasma, a continuous increase in the intensity of mass 18 (water vapour) was observed (Figure 4). This effect is explained by the heating of all materials facing the plasma and consequent temperature-stimulated desorption of water vapour. All other masses in the time range from minute 5 to minute 7 decreased with time. The masses 16, 26, and 28 were attributed to methane, acetylene and ethylene, respectively, rather than other molecules. As already mentioned, the reactor was tightly sealed, thus mass 28 could not be attributed to N_2 molecules, let alone Si atoms (the vapour pressure of Si was negligible at room or slightly elevated temperature).

Argon was introduced into the plasma reactor for 2 min only. When the Ar valve was closed and the discharge was turned off, mass 40 decreased almost instantly. Features other than mass 18 also vanished. One conclusion that could be drawn on the basis of this observation is that measurable etching of the polymer samples occurred inside the reactor under plasma conditions. Although the power density was rather small (in the case of Figure 4, the power was 3.5 kW), it was large enough to cause etching and thus formation of molecules that were not observed during pumping down and before the plasma was ignited.

Once the discharge was turned off, the mass spectrum showed only water vapour (mass 18). During operation of the diffusion pumps, the RGA signal of mass 18 decreased by almost an order of magnitude (note the logarithmic scale of the y-axis in Figure 4). Once the cold trap was open, the intensity of the RGA peak decreased by another order of magnitude (at 13 min in Figure 4), which was correlated with the instant pressure drop indicated in Figure 3.

After 14 min, the cold trap and diffusion pumps were hermetically separated from the reactor, HMDSO was introduced, and the discharge was turned on again. As a result, the RGA spectra became extremely rich (Figure S2). There were features corresponding to masses 2, 16, 18, 26, 28, 45, 52, 59, 66, 73, etc. The appearance of the peaks at large masses was attributed to various fragments of the HMDSO molecule. The precursor was thus partially decomposed both in the plasma reactor and in the RGA ionisation chamber. Any detailed analyses of RGA spectra was impossible since any heavy mass could be attributed to various molecules. The richness of the spectrum only indicated complex gas-phase chemistry. The discharge was turned off twice during deposition of the protective coating to prevent formation of dusty plasma. As explained in the Introduction, radicals tended to interact with each other in the gas phase, forming clusters that were negatively charged in plasma, thus could not escape to the surfaces that were also negative against plasma due to differences in mobility between electrons and ions. The net charge flow on the surface of any object facing plasma should be zero under steady conditions, thus the plasma should be positive against the walls to compensate for the very large electron mobility compared to ion mobility.

Mass spectrometry was therefore useful for monitoring the evolution of various molecules that were either created in gaseous plasma and stable enough to enter the RGA or created in the ionisation chamber of the RGA itself. An insight into the rather short-lived radicals that were created in the plasma was, however, obtainable from Figure 5, which presents optical spectra arising from the plasma during deposition.

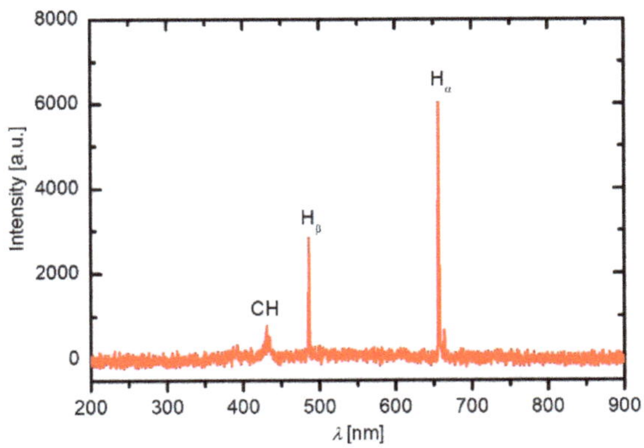

Figure 5. An optical spectrum of gaseous plasma during the deposition phase.

Figure 5 presents a typical optical spectrum during introduction of HMDSO (from minute 15 onward). Unlike the mass spectra in Figure 4, the optical spectrum in Figure 5 is extremely simple. One can observe lines that corresponded to radiative transitions from excited hydrogen atoms to the first excited state—the Balmer series, where the most intensive emission lines were the alpha and beta lines at 656 and 486 nm, respectively. There was also a peak corresponding to the CH radical. This observation was consistent with the mass spectrometry results (Figure 4), which indicate the presence of a variety of molecular fragments. Optical spectroscopy cannot reveal many-atom molecules since the radiative transitions from such molecules form a continuum. Nevertheless, the appearance of the CH peak at 431 nm revealed fragmentation of the original precursor to CH_x radicals under plasma conditions. This result is consistent with the generally accepted theory that weakly ionised plasma causes decomposition of HMDSO by subtracting methyl radicals. The large intensity of radiation arising from H atoms can be attributed to either dissociation of water vapour or subtraction of H atoms from the original precursor. The oxygen line at 777 nm was absent, which could be explained by either the poor density of O atoms in the deposition phase or low electron temperature. Both explanations are feasible, as became evident by monitoring the properties of the films deposited on sample surfaces upon performing PECVD with HMDSO.

Figure 6 presents a typical XPS depth profile of a polymer sample characterised after completing the entire production cycle, i.e., activation, metallisation and deposition of the protective coating. The substrate was composed of carbon only (XPS is not sensitive to hydrogen). There was a broad interface between pure carbon in the rightmost part of the depth profile and aluminium. The rather large concentration of oxygen at this interface was explained by successful functionalisation of the polymer surface during the activation phase (minute 5 to minute 7 in Figure 4). This functionalisation could not occur due to any interaction of argon or water vapour with the surface of the polymer material. Formation of the oxygen-rich interface between the polymer substrate and aluminium film was obviously due to plasma conditions, but a detailed explanation is not straightforward, as one should take into account various mechanisms.

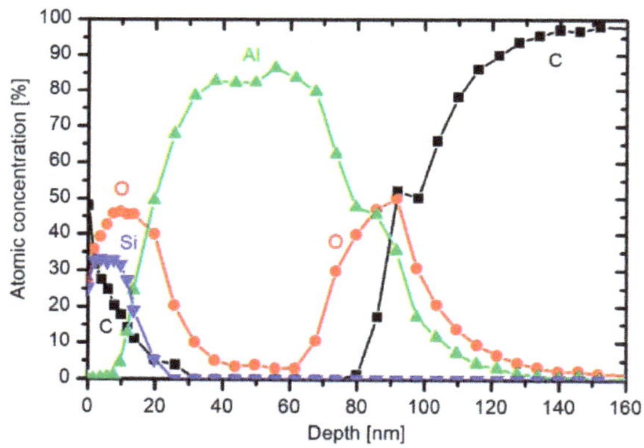

Figure 6. The XPS depth profile of the protective coating on a polymer substrate.

One method for functionalisation of a polymer surface with oxygen functional groups is irradiation with hard UV radiation and subsequent exposition to an oxygen-containing gas. A review of such a method is published recently [25]. Radiation causes bond breaking in the surface layer of the polymer, thus increasing the affinity for oxidation. The radiation in our case arose predominantly from hydrogen. There were two H lines in the optical spectrum marked as H_α and H_β arising from transitions from the second and third excited state, respectively, to the first excited state. The transitions to the ground state (which appear in the hard UV range of spectrum) were not visible, but they were much more intense than the transitions to the first excited state. According to Fanz et al. [26], the transitions to the ground state are at least an order of magnitude more intensive than those to the first excited state. The radiation arising from the transition from the first H-atom excited state to the ground state appeared at a photon energy just over 10 eV, and such photons were effectively absorbed by the surface film of the polymer, thus breaking bonds. The dangling bonds then interacted with the water vapour, which was, according to Figure 4, the major component of the residual atmosphere. The water partial pressure could not be determined accurately from the measured ion currents, but, in a rough approximation, we could account for the pressure inside the reactor before introducing argon into the system. Figure 3 reveals that the pressure was approximately 1 Pa at this stage of the process. At a pressure of 1 Pa, the density of water molecules in the reactor was as high as 10^{20} m^{-3}, which was calculated using a general gas equation:

$$p = n \times \mathrm{k} \times T, \tag{1}$$

where p is the pressure, n is the density of molecules, k is the Boltzmann constant and T is the absolute temperature. The flux of gas molecules onto the surface is

$$j = \frac{1}{4} \times n \times \langle v \rangle, \tag{2}$$

where $<v>$ is the average velocity of thermal motion (approximately 600 m/s for H_2O molecules). If all molecules interacted with dangling bonds on the polymer surface, a monolayer of oxygen would form in less than a second. Therefore, there were enough oxidative species to ensure rich functionalisation of the polymer surface prior to deposition of the aluminium film, even though the activation was performed under vacuum condition.

An alternative technique for functionalisation of a polymer surface is dissociation of water molecules into H and OH radicals. This technique is particularly useful for plasma activation of materials that release a large amount of water vapour, for example, textiles [27]. In fact, water vapour is always present in low-pressure reactors and significantly influences the surface finish of plasma-treated polymers [28]. OH radicals are renowned oxidants and interact with the polymer surface to form polar functional groups. Evidence for H_2O dissociation under plasma conditions is shown in Figure 4. The mass spectra were rich in H_2 molecules, and their appearance was attributed to the dissociation of water molecules and subsequent association of H radicals into H_2 molecules. The dissociation may occur by either electron impact or interaction with a metastable Ar atom. Therefore, there were at least two mechanisms that ensured successful formation of oxygen-rich functional groups on the surface of the polymer: (1) formation of dangling bonds on the polymer surface due to irradiation with hard UV radiation and oxidation of this modified polymer with water vapour; and (2) dissociation of water molecules under plasma conditions and interaction of OH radicals with the polymer surface. Unfortunately, the techniques adopted in this study did not allow differentiation between the mechanisms, thus it was not possible to state the prevailing one. Regardless, the high concentration of oxygen at the interface between the polymer substrate and the aluminium film, as evidenced by the XPS depth profile (Figure 6) was explained by the presence of water vapour in the plasma reactor during the activation step.

The broadness of the interface was probably not due to penetration of oxygen deep into the polymer substrate but rather to its roughness. As mentioned above, the industrial polymer samples exhibit a roughness of approximately S_a = 5 nm (S_q = 6.5 nm), which is reflected in Figure 6 as the broad interface. The oxygen-rich layer spans between approximately 70 and 120 nm in Figure 6. On top of this layer, there is a film of aluminium. The aluminium film is almost free from other elements due to the rapid deposition. Nevertheless, one can observe a measurable amount of oxygen (perhaps 5 at.%) in the aluminium film in Figure 6, which was due to the water vapour-rich residual atmosphere. The water molecules absorb on the freshly deposited aluminium and form AlO_x compounds. If the deposition occurred in a very short time or if the water concentration were negligible during aluminium deposition, the Al film would have been free from oxygen. These conditions, however, cannot be ensured in industrial reactors.

On top of the Al film in Figure 6, there is another layer rich in oxygen followed by a thin film of the protective coating. Again, the broad interface was attributed to the sample roughness. The rather large amount of oxygen in this interface was attributed to the fact that the deposition using HMDSO does not start immediately after completing Al deposition, so there is enough time for the formation of an oxide film on the freshly deposited aluminium surface (see the discussion on molecular fluxes above). Figure 7 shows the stack of Al 2*p* spectra obtained during XPS depth profiling of the layered structure of the sample. It can be seen that the spectra from the first Al-layer interface had a peak at a binding energy of 74.0 eV, meaning that Al-oxide was formed on the top of the Al-layer. The Al 2*p* spectra from the central part of the Al layer had a peak at a binding energy 71.8 eV related with Al in its metallic form. At the interface between the metallic Al and the polymer substrate, the Al 2*p* spectra again had peak at a binding energy of 74.0 eV, indicating that Al-oxide was formed in the first stage of deposition of the Al-film. The results presented in Figure 7 therefore clearly confirm partial oxidation of the Al film at both interfaces. This thin film of oxide on the surface of the aluminium coating was beneficial for adhesion of the protective coating since the surface energy (and thus wettability) of oxides is generally higher than that of pure metals.

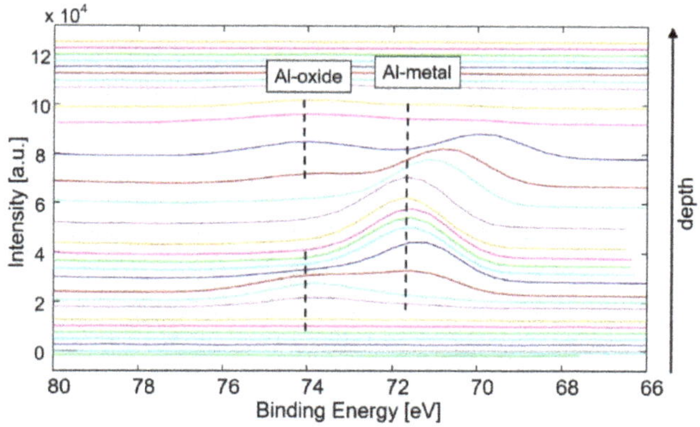

Figure 7. Al 2*p* spectra obtained during XPS depth profiling of the deposited sample.

The uppermost protective coating in Figure 6 contains silicon, oxygen and carbon. This observation was explained by partial dissociation of HMDSO under plasma conditions. Radicals were formed from the precursor, and the molecular fragments adsorbed onto the substrates, forming a layer composed of Si, C, H and O. The radicalisation led to formation of the CH_x and H radicals, wich was evident from the optical spectrum shown in Figure 5. The mass spectra (Figure 4), however, revealed very rich plasma chemistry upon depositing the protective coating on freshly metallised polymer substrates. Since water vapour was present during the PECVD process (see Figure 4), the OH radicals might interact chemically with the freshly adsorbed layer of precursor radicals, forming SiO_x, as explained above for the case of surface activation of the polymer substrate and oxidation of the topmost aluminium film.

The depth profile in Figure 6 shows only a rough composition of the silicon-rich film and does not reveal much about the structure. To obtain insight into the structure of the protective coating, we also acquired high-resolution XPS spectra and performed SIMS characterisation. Figure 8 shows high-resolution spectra acquired on the surface of a sample coated with the protective coating. The selected range of binding energies corresponded to silicon, oxygen and carbon. The silicon XPS peaks were at a binding energy of 102.3 eV. This binding energy was representative of C–Si–O bonds. The absence of a peak at 103.5 eV revealed that there was practically no SiO_2 molecules; rather, the topmost layer was a plasma-polymerised film of C–Si–O, which should offer good corrosion protection to the Al-metallisation layer [29]. As mentioned above and elaborated by Prykril et al. [8], the carbon-free coating obtained by PECVD using HMDSO was extremely fragile and tended to crack. The polymer-like coating with an appropriate structure, as revealed in this work, was more flexible than the carbon-free coating and is thus useful in many applications. Figure 8 also shows the high-resolution oxygen O 1*s* spectrum with peaks at 532.7 eV. This binding energy is often related to C–Si–O bonds, thus the result is consistent with the position of the Si peak. Finally, the carbon C 1*s* XPS spectrum is also shown in Figure 8. The binding energy of this peak was at 284.8 eV, which was assigned to carbon bonded with hydrogen, silicon or another carbon rather than to oxidised carbon such as O–C=O, C=O, or C–O. The high-resolution XPS spectra therefore indicated a structure practically free from silica and carbon atoms bonded to silicon rather than oxygen.

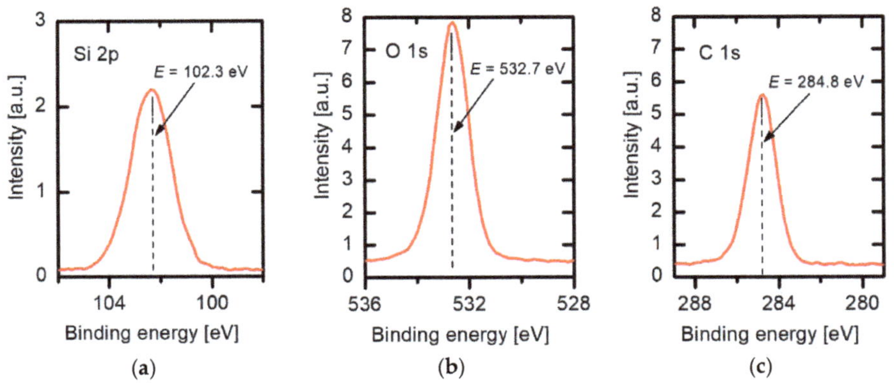

Figure 8. High energy resolution XPS spectra obtained on the surface of the deposited sample of: (**a**) Si 2*p*; (**b**) O 1*s*; and (**c**) C 1*s*.

Additional information about the structure of the deposited protective coating is available from an alternative technique for surface characterisation-ToF-SIMS. The spectrum of positively charged secondary ions emitted from the surface of a sample irradiated with energetic bismuth clusters is shown in Figure 9 in the mass range of 1–200 amu. The main peaks in this spectrum were at 27.97 (Si^+), 42.99 ($SiCH_3^+$), 59.03 ($SiC_2H_7^+$), 73.05 ($SiC_3H_9^+$) and 147.05 amu ($Si_2C_5H_{15}^+O$). These peaks were actually clusters of the Si–C–H–O compound that forms the uppermost protective layer. The peaks were typical of polydimethylsiloxane [30]. The SIMS spectrum shown in Figure 9 is therefore consistent with the XPS spectra shown in Figure 8. The high concentration of hydrogen in the secondary ion clusters also indicated the polymer-like structure of the deposited protective coating.

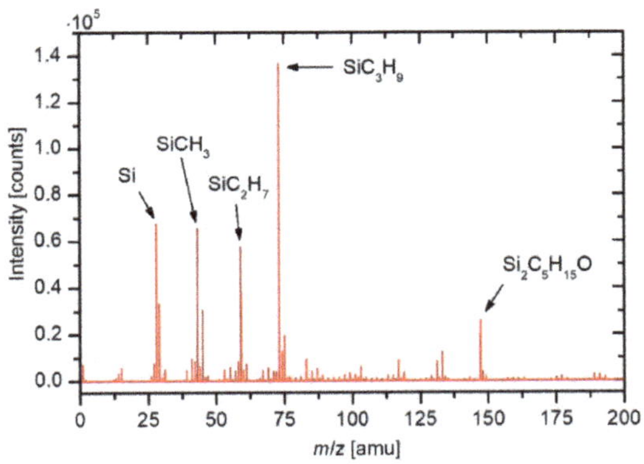

Figure 9. ToF-SIMS spectrum of positive secondary ions emitted from the surface of the deposited sample.

A fundamental discrepancy between the processes of plasma activation and deposition of the protective coating might be observed. As explained above, the gaseous plasma was suitable for formation of oxygen-rich functional groups on the polymer surface during the activation process (5–7 min in Figures 3 and 4), however no measurable oxidation of the protective polymer coating was revealed from the XPS and SIMS results, as shown in Figures 6–9. This fundamental discrepancy was explained by the various gas-phase and surface reactions involved. The activation process was performed using argon plasma and taking advantage of the presence of water vapour to ensure appropriate surface activation of the polymer substrate. Although the power density was low, the irradiation with bond-breaking UV radiation and oxidation with OH or H_2O molecules were sufficient to functionalise the substrate with polar groups rich in oxygen. During the plasma polymerisation step, however, HMDSO molecules were the predominant gaseous species. Compared to Ar molecules, HMDSO molecules had numerous channels for the loss of electron energy upon collision. The numerous dissociative collisions effectively lowered the electron temperature in HMDSO compared to Ar, thus the electrons were not energetic enough to atomise the molecules. The majority of the available electron energy in the HMDSO plasma was therefore spent on only partial radicalisation of these molecules. As a result, the main backbone of HMDSO was preserved, thus the protective coating grew from molecular fragments rather than atoms. Silicon in HMDSO therefore remained bonded to oxygen and carbon, and this material exhibited much less affinity for oxidation by OH radicals than the polymer substrate, let alone freshly deposited aluminium. Furthermore, as the electron energy was spent preferentially on radicalisation of HMDSO, the water vapour was not dissociated to reactive OH radicals as much as it was in argon plasma, thus the available flux of this highly oxidative radical was not large enough to cause complete oxidation of the Si–C–H–O coating and formation of SiO_x within the protective coating.

4. Conclusions

The combination of various techniques for plasma and surface characterisation adopted in this work provided insight into the mechanisms involved both in the gas phase and on the surfaces of materials facing plasma upon deposition of an appropriate protective coating onto polymer substrates in a large industrial plasma reactor. The reactor configuration enabled appropriate gas composition during all three production steps: surface functionalisation, aluminization and deposition of a protective coating resembling polydimethylsiloxane. The activation process was performed with argon plasma. The lack of electron energy loss mechanisms in argon plasma resulted in preferential dissociation of water vapour molecules present in the residual atmosphere. The resultant H atoms were excited to radiative states, providing UV radiation that caused bond breaking in the polymer surface. The OH radicals enhanced the oxidation of the polymer surface and thus the formation of oxygen-rich functional groups that endowed the polymer substrates with appropriate surface energy to ensure good adhesion of the aluminium coating. In the phase of plasma polymerisation of the protective coating, however, the electron energy was spent preferentially on partial dissociation of the HMDSO precursor. The nature of the deposited film and lack of highly reactive oxygen-containing radicals enabled preservation of the C–Si–O bonds and thus formation of a coating resembling polydimethylsiloxane rather than more extensive oxidation of the protective coating and consequent formation of SiO_x clusters in the film, which was observed by several other authors upon intentionally adding oxygen to HMDSO upon plasma deposition of protective coatings [6,8,10–12].

Supplementary Materials: The following are available online at http://www.mdpi.com/2079-6412/9/4/234/s1, Figure S1: One of the AFM images of the untreated sample. Measurements with AFM were performed on 3 samples at different places. The average roughness was around S_a = 5 nm (S_q = 6.5 nm), Figure S2: Mass spectrometry of gaseous molecules of higher masses during all phases of the production process.

Author Contributions: Conceptualisation, Ž.G., M.M. and R.Z.; Methodology, A.V. and J.K.; Validation, G.P. and A.V.; Formal Analysis, R.Z. and J.K.; Investigation, R.Z., J.K. and Ž.G.; Resources, Ž.G., R.Z. and J.K.; Data Curation, J.K., R.Z. and G.P.; Writing—Original Draft Preparation, Ž.G. and M.M.; Writing—Review and Editing, Ž.G., M.M., A.V. and R.Z.; Visualisation, R.Z.; Supervision, Ž.G. and M.M.; Project Administration, A.V. and G.P.; Funding Acquisition, M.M.

Funding: This research was funded by the Slovenian Research Agency (research core funding No. P2-0082-Thin film structures and plasma surface engineering, as well as project No. L2-8179-Evaluation of the range of plasma parameters suitable for nanostructuring of polymers on industrial scale).

Acknowledgments: The authors thank the Elvez Company for allowing them to use one of their large industrial plasma reactor.

Conflicts of Interest: The authors declare no conflict of interest.

References

1. Hegemann, D.; Nisol, B.; Watson, S.; Wertheimer, M.R. Energy conversion efficiency in low- and atmospheric-pressure plasma polymerization processes, Part II: HMDSO. *Plasma Chem. Plasma Process.* **2017**, *37*, 257–271. [CrossRef]
2. Zhang, J.; Ji, X.; Bao, J.; Zheng, X. Deposition and characterization of Si-doped diamond films using tetraethoxysilane onto a WC-Co substrate. *Coatings* **2016**, *6*, 39. [CrossRef]
3. Esbayou, M.; Bentiss, F.; Casetta, M.; Nyassi, A.; Jama, C. Optimization of cold plasma process parameters for organosilicon films deposition on carbon steel: Study of the surface pretreatment effect on corrosion protection performance in 3 wt % NaCl medium. *J. Alloy. Compd.* **2018**, *758*, 148–161. [CrossRef]
4. Lien, S.Y.; Lin, Y.S.; Cho, Y.S.; Wuu, D.S. Performance of flexible photovoltaic modules encapsulated by silicon oxide/organic silicon stacked layers. *IEEE Trans. Electron Devices* **2016**, *63*, 1615–1620. [CrossRef]
5. Vautrin-Ul, C.; Boisse-Laporte, C.; Benissad, N.; Chausse, A.; Leprince, P.; Messina, R. Plasma-polymerized coatings using HMDSO precursor for iron protection. *Prog. Org. Coat.* **2000**, *38*, 9–15. [CrossRef]
6. Chang, Y.; Liao, W.; Wei, H.; Liao, Y.; Chen, H.; Kuo, C. Fabrication of transparent very thin SiO_x doped diamond-like carbon films on a glass substrate. *Coatings* **2018**, *8*, 240. [CrossRef]
7. Mitschker, F.; Schücke, L.; Hoppe, C.; Jaritz, M.; Dahlmann, R.; de los Arcos, T.; Hopmann, C.; Grundmeier, G.; Awakowicz, P. Comparative study on the deposition of silicon oxide permeation barrier coatings for polymers using hexamethyldisilazane (HMDSN) and hexamethyldisiloxane (HMDSO). *J. Phys. D Appl. Phys.* **2018**, *51*, 235201. [CrossRef]
8. Prikryl, R.; Otrisal, P.; Obsel, V.; Svorc, L.; Karkalic, R.; Buk, J. Protective properties of a microstructure composed of barrier nanostructured organics and SiO_x layers deposited on a polymer matrix. *Nanomaterials* **2018**, *8*, 679. [CrossRef]
9. Dos Reis, M.C.; Silva, V.R.M.; Sgura, R.; da Cruz, N.C.; Rangel, E.C.; Medeiros, I.S. Surface characteristics and optical properties of plasma deposited films on indirect aesthetic restorative dental materials. *Surf. Coat. Technol.* **2018**, *348*, 55–63. [CrossRef]
10. Brochhagen, M.; Chur, S.; Layes, V.; Böke, M.; Benedikt, J. Deposition of SiO_x coatings by inductively coupled plasma: Effect of pulsed hexamethyldisiloxan flow. *Plasma Process. Polym.* **2018**, *15*, 1700186. [CrossRef]
11. Pandiyaraj, K.N.; Kumar, A.A.; Ramkumar, M.C.; Kumar, S.U.; Gopinath, P.; Cools, P.; De Geyter, N.; Morent, R.; Bah, M.; Shah, S.I.; et al. Effect of processing parameters on the deposition of SiO_x-like coatings on the surface of polypropylene films using glow discharge plasma assisted polymerization for tissue engineering applications. *Vacuum* **2017**, *143*, 412–422. [CrossRef]
12. Hoppe, C.; Mitschker, F.; Giner, I.; de los Arcos, T.; Awakowicz, P.; Grundmeier, G. Influence of organic surface chemistry on the nucleation of plasma deposited SiO_x films. *J. Phys. D Appl. Phys.* **2017**, *50*, 204002. [CrossRef]
13. Loffhagen, D.; Becker, M.M.; Czerny, A.K.; Philipp, J.; Klages, C.P. Impact of hexamethyldisiloxane admixtures on the discharge characteristics of a dielectric barrier discharge in argon for thin film deposition. *Contrib. Plasma Phys.* **2018**, *58*, 337–352. [CrossRef]
14. Pavlinak, D.; Galmiz, O.; Zemanek, M.; Cernak, M. Design and evaluation of plasma polymer deposition on hollow objects by electrical plasma generated from the liquid surface. *Plasma Process. Polym.* **2018**, *15*, 1700183. [CrossRef]
15. Cui, X.; Zhu, G.; Pan, Y.; Zhang, Y. Preparation and anticorrosion properties of Polydimethylsiloxane (PDMS) coating. In Proceedings of the 4th Annual International Conference on Material Engineering and Application (ICMEA 2017), Wuhan, China, 15–17 December 2017.
16. Xue, C.H.; Bai, X.; Jia, S.T. Robust, self-healing superhydrophobic fabrics prepared by one-step coating of PDMS and octadecylamine. *Sci. Rep.* **2016**, *6*, 27262. [CrossRef]
17. Park, J.Y.; Yoo, S.J.; Lee, E.J.; Lee, D.H.; Kim, J.Y.; Lee, S.H. Increased poly(dimethylsiloxane) stiffness improves viability and morphology of mouse fibroblast cells. *BioChip J.* **2010**, *4*, 230–236. [CrossRef]

18. Palchesko, R.N.; Zhang, L.; Sun, Y.; Feinberg, A.W. Development of Polydimethylsiloxane substrates with tunable elastic modulus to study cell mechanobiology in muscle and nerve. *PLoS ONE* **2012**, *7*, e51499. [CrossRef]
19. Chuah, Y.J.; Wu, Y.; Yang, Z.; Lee, E.H.; Kang, Y. Geometrical, topographical, and mechanical cues on stem cell fate in a micro-/nano-environment. In *Encyclopedia of Microfluidics and Nanofluidics*; Li, D., Ed.; Springer: Boston, MA, USA, 2015; pp. 1277–1284.
20. Huh, D.; Hamilton, G.A.; Ingber, D.E. From 3D cell culture to organs-on-chips. *Trends Cell Biol.* **2011**, *21*, 745–754. [CrossRef] [PubMed]
21. Menon, N.V.; Chuah, Y.J.; Cao, B.; Lim, M.; Kang, Y. A microfluidic co-culture system to monitor tumor-stromal interactions on a chip. *Biomicrofluidics* **2014**, *8*, 064118. [CrossRef] [PubMed]
22. Eduok, U.; Faye, O.; Szpunar, J. Recent developments and applications of protective silicone coatings: A review of PDMS functional materials. *Prog. Org. Coat.* **2017**, *111*, 124–163. [CrossRef]
23. Zaplotnik, R.; Vesel, A.; Mozetič, M. Atomic oxygen and hydrogen loss coefficient on functionalized polyethylene terephthalate, polystyrene, and polytetrafluoroethylene polymers. *Plasma Process. Polym.* **2018**, *15*, 1800021. [CrossRef]
24. Mozetic, M.; Drobnic, M.; Zalar, A. Recombination of neutral hydrogen atoms on AISI 304 stainless steel surface. *Appl. Surf. Sci.* **1999**, *144*, 399–403. [CrossRef]
25. Zhang, Y.; Ishikawa, K.; Mozetič, M.; Tsutsumi, T.; Kondo, H.; Sekine, M.; Hori, M. Polyethylene terephthalate (PET) surface modification by VUV and neutral active species in remote oxygen or hydrogen plasmas. *Plasma Process. Polym.* **2019**, 1800175. [CrossRef]
26. Fantz, U.; Briefi, S.; Rauner, D.; Wünderlich, D. Quantification of the VUV radiation in low pressure hydrogen and nitrogen plasmas. *Plasma Sources Sci. Technol.* **2016**, *25*, 045006. [CrossRef]
27. Gorjanc, M.; Bukošek, V.; Gorenšek, M.; Vesel, A. The influence of water vapor plasma treatment on specific properties of bleached and mercerized cotton fabric. *Text. Res. J.* **2010**, *80*, 557–567. [CrossRef]
28. Gorjanc, M.; Jazbec, K.; Šala, M.; Zaplotnik, R.; Vesel, A.; Mozetič, M. Creating cellulose fibres with excellent UV protective properties using moist CF_4 plasma and ZnO nanoparticles. *Cellulose* **2014**, *21*, 3007–3021. [CrossRef]
29. Grundmeier, G.; Thiemann, P.; Carpentier, J.; Barranco, V. Tailored thin plasma polymers for the corrosion protection of metals. *Surf. Coat. Technol.* **2003**, *174*, 996–1001. [CrossRef]
30. Alexander, M.R.; Jones, F.R.; Short, R.D. Mass spectral investigation of the radio-frequency plasma deposition of hexamethyldisiloxane. *J. Phys. Chem. B* **1997**, *101*, 3614–3619. [CrossRef]

coatings

MDPI

Article

Influence of Laser Colour Marking on the Corrosion Properties of Low Alloyed Ti

Tadeja Kosec [1,*], Andraž Legat [1], Janez Kovač [2] and Damjan Klobčar [3]

1 Laboratory for Metals, Corrosion and Anticorrosion Protection, Slovenian National Building and Civil Engineering Institute, Dimičeva 12, SI-1000 Ljubljana, Slovenia; andraz.legat@zag.si
2 Department for Optoelectronics, Jožef Stefan Institute, Jamova 39, SI-1000 Ljubljana, Slovenia; janez.kovac@ijs.si
3 Laboratory for Welding LAVAR, Faculty of Mechanical Engineering, University of Ljubljana, Aškerčeva 6, SI-1000 Ljubljana, Slovenia; damjan.klobcar@fs.uni-lj.si
* Correspondence: tadeja.kosec@zag.si; Tel.: +386-1-280-4547

Received: 7 May 2019; Accepted: 6 June 2019; Published: 9 June 2019

Abstract: In the field of surface treatment, laser colour marking can be used to produce coloured marks on the surfaces of metals. Laser colour markings can be applied to various materials, but on titanium alloys a wide spectra of vivid colours can be achieved. This study presents an analysis of the corrosion properties of laser treated surfaces that were exposed to aggressive environments. Different samples were prepared with laser light of various power intensities and processing speeds. The samples were prepared on low alloyed Ti. Electrochemical, spectroscopic and microstructural analyses were conducted in order to study the properties of the laser treated surfaces. Corrosion testing showed different effects of laser power and production speed on the properties of the laser treated surfaces. It was shown that a high intensity and slow processing rate affect the surfaces by forming oxides that are relatively stable in a corrosive environment of 0.1 M NaCl. Spectroscopic investigations including scanning electron microscopy (SEM) and X-ray photoelectron spectroscopy (XPS) analyses showed the differences in chemical structure of the surface layer formed after laser treatment. Similarly, microstructural investigations showed different effects on the surface and sub-surface layer of the laser treated samples.

Keywords: Ti alloy; laser color creation; microstructure; passive film; X-ray photoelectron spectroscopy (XPS); corrosion

1. Introduction

The aims of this study were to study the properties of laser treated Ti surfaces and to study the effect on corrosion resistivity in a 0.1 M NaCl solution.

The marking of surfaces is widely used in industry in order to distinguish industrial products from each other or to provide certain information to the customers through bar codes or similar marks [1,2]. Some producers use marking to distinguish their original parts. Marking can be completed on plastics and different metals [3–12]. On certain metals and/or alloys, like stainless steels, nickel alloys, chrome plated steel, aluminium and titanium alloys, marks of different colours can be obtained [5–12]. This effect is generally used in the automotive, electronics, telecommunications and pharmaceutical industries for general consumer goods and high-end products, for decorative purpose and advertising, in the production of jewellery and even art [12–20].

There are various technologies used to produce markings, including printing, plasma treatment [11,21], electrochemical treatment, magnetron sputtering, micro-arc oxidation [22,23], anodising [1,8,24], cold spray [17] and laser induced oxidation [1,2,9,25]. The most important advantages of laser induced oxidation are that the process is non-contact and that the marking of smaller surfaces can be made fast at relatively low cost. The produced markings are very sharp, reproducible and precise. Another benefit is that the process enables flexible change of marking colours on the fly with respect to laser processing parameters [6,13–16].

Titanium and its alloys have several specific properties, like good strength to weight ratio, high strength, good chemical stability, high temperature stability, biocompatibility, corrosion resistance and resistance to fatigue and wear, and can be heat treated or surface treated in order to achieve particular hardness and strength [11,26].

Laser induced oxidation of metals is a gaseous high-temperature corrosion process, where thin layers of oxides or other more complex oxidation formations are formed. With this process, different colours can be produced on the titanium surface. During laser oxidation, titanium oxides are formed at the surface if the marking is carried out in an oxidative atmosphere. Titanium oxides are not expensive, biocompatible, non-toxic and mechanically and chemically stable [16]. On the surfaces, there are thin layers present in the form of anatase, brookite, rutile or in the amorphous phase [6,13,16]. In the white light spectrum, they are highly transparent and exhibit a wide spectrum of vivid colours as a result of the interference between white light and the borders between different phases. The obtained colours depend on the thickness of the oxide layer, refractive index and the interference between light and incidence angle [6,13,15,27]. There are many laser parameters, such as pulse duration, pulse repetition rate, pulse energy, focal position of laser beam and processing parameters, such as scanning speed, hatching distance, laser fluence and scanning strategy, that change the properties of the produced compounds and the thickness of the oxide layers. Among these parameters, the produced colours, i.e., layer thickness and its composition, are directly affected by scanning speed, while the reproducibility of colours is highly sensitive to hatching distance at constant laser parameters. Researchers have also noted that other parameters, such as material surface temperature and laser fluence, have an influence on the produced colour but are hard to control if the size and shape of the product is changing [6,13–16,28].

The reactions and their product compositions are based on thermodynamic data and kinetic coefficients. Since laser induced oxidation is usually done in an air atmosphere (mixture of gases), the processes are non-isothermal and thus the oxidation results are usually very complex. Different authors have reported that observed layers consisted of different titanium oxides, namely, TiO, TiO_2, Ti_2O and Ti_2O_3 [6,7,12,16]. Adams et al. [29] performed the analysis using nanosecond fibre laser marking of titanium grade 2. They reported that the surface coating is made of three layers, TiO_2 (top), TiO (middle) and a bottom layer, which is an inhomogeneous mixture of Ti, O and N, from which only the two top layers are optically transparent. Researchers have proposed two- and three-layer models of coatings, where the three-layer model is made of (TiO_2/TiO/Ti_6O + TiO_xN_{1-x}) [6,16]. Some authors also pointed out a TiN phase and its influence on the mechanical properties of the surface [13]. Antonzcak et al. [16] studied the presence of nitrogen compounds and the existence of nonstoichiometric factors ($TiN_{0.26}$, $TiN_{0.22}O_{0.78}$ and $TiO_{0.892}$) using X-ray photoelectron spectroscopy (XPS) and X-ray diffraction (XRD) analyses. They showed that the top layer of the coating is not made of pure TiO_2 [16].

Perez del Pino et al. [7] reported that they obtained a coating layer thickness of between 0.1 and 10 μm, and that the thickness depends on the laser fluence. Adams et al. [29] showed that the visually transparent part of the coating layer is made of TiO_2 and TiO, and its thickness is between 10 and 120 nm.

Analysis of the morphology showed that rapid heating and cooling of the surface during laser marking causes high temperature gradients, which cause the formation of a channel micro-crack network [13]. The micro-cracks have an influence on the diffusion rate of oxygen, they change the reaction speed and composition of the produced layer and its effect on corrosion resistance [15,30], and could cause the delamination problems if laser marking is done with inappropriate fluence. However, the adhesion between the oxide layer and the titanium surface is good, as stated by Akman et al. [13]. The scanning speed also has an influence on the grain size, which becomes larger if the scanning speed is smaller. Akman et al. [13] also showed that laser surface treatment has an influence on the scratch hardness, which becomes higher if the scanning speed is smaller. Antonzak et al. [16] showed that the values of the roughness parameters become smaller if the laser fluence increases. Such a layer hardens, thus as reported, the durability of titanium is reduced [31]. So far, corrosion testing has been performed on martensitic and austenitic stainless steels in Ringers' solution in order to study the degradation of identification marks on surgical tools and prostheses [32].

In this study, different microstructural and spectroscopic investigations have been used to evaluate the effect of laser on Ti-based materials. XPS was used to study the surface properties of treated materials and the influence of the laser power and the scan rate of the laser. Electrochemical techniques, such as potentiodynamic and electrochemical impedance spectroscopy were used in order to study electrochemical properties of newly formed oxide films, as well as their stability in slightly saline solution simulating road conditions.

2. Materials and Methods

2.1. Preparation of Samples

Samples from low alloyed Ti were prepared in a shape of discs of $\Phi = 16$ mm, cut from 1.5-mm-thick foil. Composition is given in Table 1. The studied Ti alloy is suited for automotive and other industrial applications [33]. It exhibits superior oxidation resistance at elevated temperatures compared to other commercially pure Ti [33].

Table 1. Chemical composition of low alloyed Ti, analysed by energy-dispersive X-ray spectroscopy (EDS, Inca, Oxford, UK).

Element	Si	Al	Cr	Fe	Ti
Content (wt %)	0.48	0.4	1.27	2.81	balance

Ti sheets were treated by shot-peening process using stainless steels beads (appearance of Fe and Cr). The samples were ultrasonically cleaned in ethanol for 3 min prior to laser treatment and prior to the measurements. Laser induced colour marking was done using a Speedy 400 flexx engraving machine (Trotec, Marchtrenk, Austria). The deflection of the laser beam on the desired surface of the substrate was completed using a CNC- computed numeric control) table. The colour marking was completed using a Yb glass fibre laser with a wavelength of 1064 nm, an average output power of up to 30 W, a peak power output of up to 15 kW, a pulse energy of 1 mJ, a pulse duration between 4 and 200 ns and beam quality factor M2 $\leq$ 1.5 (YLPN, IPG, Burbach, Germany-1-4x200-30-M). The constant colour marking parameters were a pulse repetition rate (PRR) at 35 kHz and defocus at -2 mm, with an air atmosphere present during marking that was done at 500 dpi. During the marking, the fibre laser power and marking speed changes according to Table 2.

Table 2. Parameters of fibre laser colour marking.

Sample	Speed [%]	Speed [mm/s]	Average Power [%]	Average Power [W]	PRR [kHz]	Observed Colour
(a) Low alloyed Ti	–	–	–	–	–	Natural grey
(b) Ti_laser_65-3	3	60	65	19.5	35	Brown
(c) Ti_laser_100-6.5	6.5	130	100	30		Yellow
(d) Ti_laser_100-4	4	80	100	30		Light blue
(e) Ti_laser_100-3	3	60	100	30		Blue

2.2. Microstructural Examination

Samples for metallographic investigation were etched in a solution of 100 mL of H_2O, 2 mL of HF and 5 mL of conc. HNO_3 for 2 min. Shortly afterwards, an optical microscopy study was conducted at different magnifications. The number of inclusions was estimated at scans with a magnification of 20×.

2.3. Electrochemical Measurements

The test solution was 0.1 M NaCl. A three-electrode corrosion cell was used, with a volume of 350 cm^3. The working electrode was embedded in a Teflon holder, and had an exposed area of 0.785 cm^2. Reference 600+ (Gamry, Warminster, PA, USA) was used for electrochemical measurements. The short electrochemical tests were performed after 1-h stabilisation at open circuit potential (OCP) and included measuring of corrosion potential (1 h), electrochemical impedance spectroscopy (after 1.33 h) followed by potentiodynamic tests. Electrochemical impedance spectroscopy (EIS) measurements at 10 rms, seven points per decade and from 65 kHz to 1 mHz were conducted at open circuit potential. At the end of short-term measurements, the potentiodynamic measurements were performed starting from −0.25 V vs. OCP, and progressing in the anodic direction up to +2.0 V at a scan rate of 1 mV/s. For long-term experiments, EIS was conducted after 24 h and then weekly (1 week, 2 weeks and 3 weeks). All potentials are reported with respect to the saturated calomel electrode (SCE) scale. At least three measurements were performed in order to fulfil the statistical requirements for electrochemical testing [34]. After estimating the mean values of the logarithmic results of corrosion resistance, the measurement that had the closest value to the mean value from the set was chosen to be presented in the graphs.

2.4. Surface Analysis

X-ray photoelectron spectroscopic (XPS) analyses were performed on a PHI-TFA XPS (TFA XPS, Physical Electronics Inc., Chanhassen, MN, USA) instrument equipped with an Al-monochromatic X-ray source. The analysis area was 400 μm in diameter and I XPSHI the analysis depth was ~5 nm, which shows very high surface sensitivity of this method. In order to obtain the subsurface chemical composition, Ar-ion sputtering was performed in addition to XPS analyses. Signals of Ti 2*p*, O 1*s*, C 1*s*, Fe 2*p*, Si 2*p* and N 1*s* were collected during XPS depth profile analyses. An Ar-ion beam of 4 keV scanning over a 3 mm × 3 mm region on the surface was employed for depth profiling. The sputtering rate was estimated to be 1 nm/min measured on a reference sample of known thickness. By XPS depth profiling, a subsurface region of ~25 nm was analysed.

Scanning electron microscopy (SEM) was performed on a low vacuum JEOL 5500 LV scanning electron microscope (JEOL, Akishima, Japan), equipped with Oxford Inca (EDX) energy dispersive spectroscopy (Oxford Instrument Analytical, Abingdon, UK), in order to examine the alloy composition by using an accelerating voltage of 20 kV.

3. Results and Discussion

3.1. Light Mictroscopy and Microstructural Examination

The optical image of the studied surfaces is presented in Figure 1.

The colours were characteristic for high-temperature titanium oxides. The surface of the low alloyed Ti was mechanically treated after rolling the sheet metal (Figure 1a) by a shot peening process. The surface appearance of the laser marked Ti surface (laser 65_3) was lightly cracked (Figure 1b), as observed in a rectangle in Figure 1b. There was a significant number of impurities observed on the surface (Figure 1b). In the case of a higher production speed of a laser (Figure 1c), the surface layer was not completely melted. Traces of the original surface and minor cracks in all directions were visible (white rectangle in Figure 1c). When the laser beam with high power passed, multiple tiny cracks were formed on the surface (depicted in white rectangles in Figure 1d,e), which extended transversely to the direction of the laser beam passage. The cracks were more intensive when the power of the laser was increased and when the processing speed was lowered. The surface of the colour marked samples looked wavy, but this was not visible in the cross section (Figure 2). Cracking is, according to the literature, related to α-Ti phase formation, which is capable of dissolving ~33 at % oxygen and 23 at % nitrogen. This is reported as an oxygen-rich metallic layer, namely, Ti(O) interstitial solid [16].

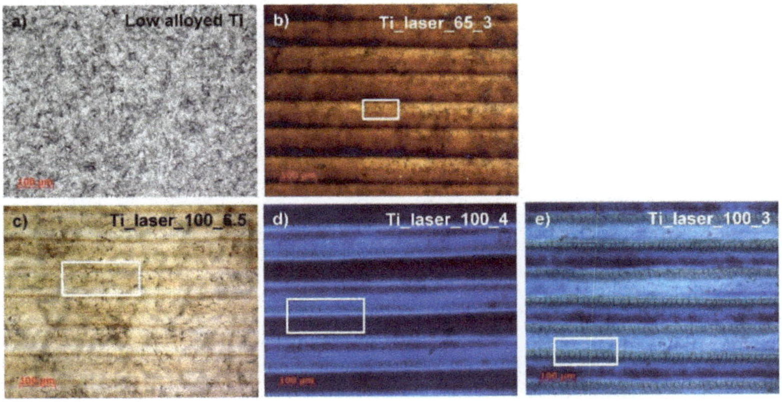

Figure 1. Prepared laser colour marking of low alloyed Ti for the samples presented in Table 1. (**a**) low alloyed Ti; (**b**) sample Ti_laser_65_3; (**c**) sample Ti_laser_100_6.5; (**d**) sample Ti_laser_100_4; (**e**) sample Ti_laser_100_3.

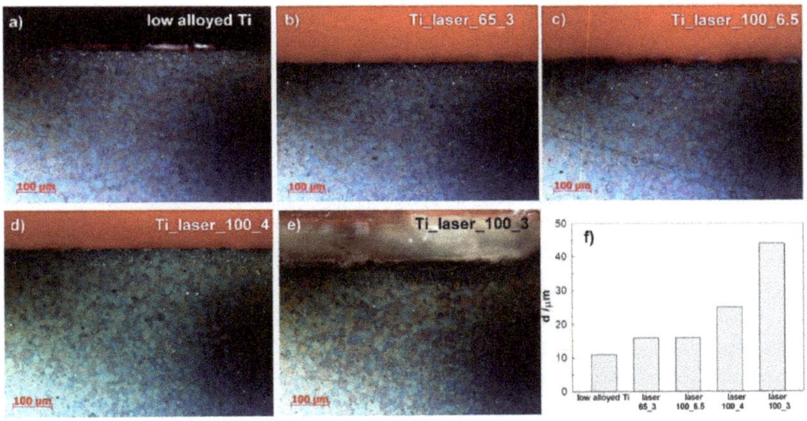

Figure 2. Microstructural investigation of laser marked surfaces in a cross-sectional view on low alloyed Ti foil (**a**), on a sample Ti_laser_65_3 (**b**), on a sample Ti_100_6.5 (**c**), on a sample Ti laser_100_4 (**d**), on a sample Ti_laser_100_3 and (**e**) the effect on the changed heat affected zone in cross section (**f**).

Laser marked surfaces were metallographically examined in the cross section of the prepared surfaces on disc electrodes. The results are presented in Figure 2.

Low alloyed Ti has a typical microstructure with crystal grains oriented in all directions equally. The mechanical treatment induced the recrystallization layer, which is observed on low alloy Ti and is approximately 11 µm thick. When the surface is laser treated, the heat, produced by the laser, affects the microstructure in the upper layers to some extent. Cross-sectional examination of laser treated surfaces showed that the intensity (power) affects the depth of the changed microstructure. In addition, the effect varies with the production speed of the laser, where the depths of the affected microstructure in µm are presented in Figure 2f. The most affected microstructure, thus, was the microstructure below a surface of low alloyed Ti treated with 100% laser power and lowest production speed (laser_100_3).

The crystal grains became smaller in the range of 1–5 µm, whereas original crystal grains at a depth of 100 to 200 µm were 10–20 µm.

3.2. Electrochemical Measurements in 0.1 M NaCl Solution

Potentiodynamic measurements were made on low alloyed Ti surfaces and laser treated surfaces after 3 h immersion in 0.1 M NaCl. The results are presented in Figure 3. The electrochemical parameters, deduced from the curves, are presented in Table 2.

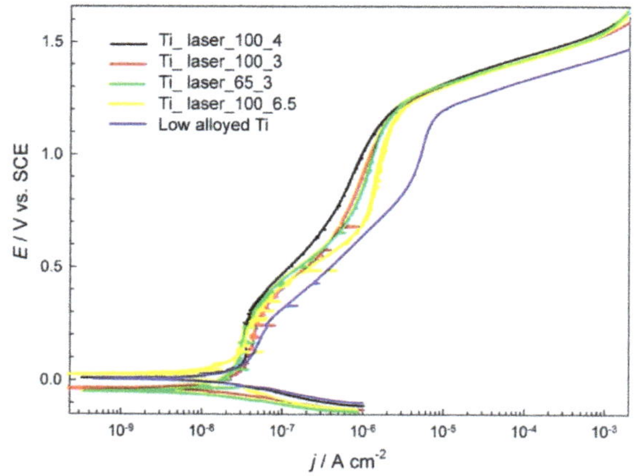

Figure 3. Potentiodynamic polarisation curves for Ti and laser treated Ti surfaces in 0.1 M NaCl solution at a scan rate of 1 mV/s.

In the case of low alloyed Ti, it is clearly shown that the Ti surface exhibits passive behaviour, as observed from low passive current densities in the passive region. The breakdown potential, E_b, for low alloyed Ti is at 1.22 V and was estimated at a potential at which transpassive region with slow and steeper increase of a current density is observed (Figure 3). The current density in the pseudo passive region for non-treated low alloyed Ti is the highest among the tested samples. Samples treated with 100% laser at the highest marking speed (6%), lowest power (65%) and 3% marking speed have the second highest current density in the passive region whereas the corrosion current density is increased. Surfaces that were treated with the highest power of 100% and a smaller marking speed exhibited the smallest corrosion current density in the anodic region. Furthermore, the smallest corrosion current density j_{corr} was measured for 100% power and 4% marking speed at 16.7 nA·cm^{-2}. The breakdown potential is not affected to a great extent. The passivity region ΔE ($E_b - E_{corr}$) in Table 3 shows that the width of a passive region is wider for laser treated surfaces.

Table 3. Electrochemical parameters from potentiodynamic curves.

Sample	E_{corr}/V	j_{corr} */nA·cm^{-2}	E_b/V	ΔE/V
Low alloyed Ti	0.011	27.0 ± 2.2	1.17	1.16
Ti Laser 100_6.5	0.030	47.9 ± 1.0	1.22	1.19
Ti Laser 100_4	0.009	16.1 ± 0.6	1.22	1.21
Ti Laser 100_3	−0.033	28.5 ± 0.5	1.22	1.25
Ti Laser 65_3	−0.044	29 ± 9	1.22	1.27

*Average value of corrosion current densities.

In Figure 4 the representative impedance measurements are presented in the form of a Nyquist and Bode plot for low alloyed Ti and laser treated Ti surface (laser_100_3) after 1.33 h immersion time during short term experiments. For short term experiments, the total impedance at lowest measured frequency, |Z|, for low alloyed Ti and surface treated Ti (sample Ti laser_100_3) was 2.01 and 1.98 MΩ·cm^2, respectively.

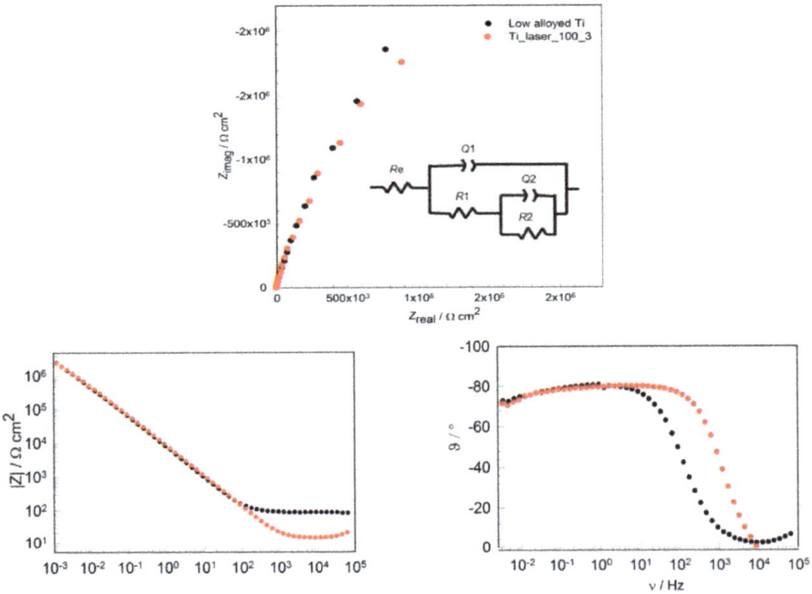

Figure 4. Nyquist and Bode plot for low alloyed Ti and laser treated surface (Ti Laser_100_3) in 0.1 M NaCl after 1.33 h immersion time at short term experiment.

The impedance spectra for short term experiments were fitted with the equivalent circuit presented in Figure 4. Equivalent circuit consists of parallel combination of resistance and capacitance elements (RQ) that are in series with R_e, electrolyte resistance. The circuit represents just one of possible equivalent circuits that are adequate to fit the impedance spectra. This model was chosen on the basis of Pan et al. to describe a bi-layer structure of oxide film on titanium in a saline environment [35]. The film, that forms on Ti based alloy, also exhibits a two layer structure with a dense inner layer of TiO_2 and porous outer layer.

The high frequency parameters R_1 and Q_1 represent the properties of the outer porous and passive film/solution interface reactions. The symbol Q signifies the possibility of a non-ideal capacitance (CPE, constant phase element). A CPE is usually used to describe non-ideal capacitive behaviour due to uneven current distribution or surface inhomogeneity providing the exponent n is close to unity. The impedance of the CPE is given by [36]:

$$Q = Z_{\mathrm{CPE}}(\omega) = [C(j\omega)^n]^{-1} \quad (1)$$

for $n = 1$, the Q element reduces to a capacitor with a capacitance C and, for $n = 0$, to a simple resistor. The values of parameter n around 0.5 indicate the diffusion process through the pores of oxide film.

The process in the low frequency range describes the capacitance of the barrier layer (Q_2) at the electrolyte/dense passive film interface. R_2 is the charge transfer resistance. The values of fitted parameters of the equivalent circuit at initial immersion time for low alloyed Ti and laser treated Ti surface are presented in Table 4.

Table 4. Values of fitted parameters of the equivalent circuit as a function of applied potential at 1 h immersion time. Units: $R[\Omega{\cdot}cm^2]$, $Q[\Omega^{-1}{\cdot}cm^{-2}s^n]$ and $C[F{\cdot}cm^{-2}]$.

Sample	R_e	Q_1	C_1	n_1	R_1	Q_2	C_2	n_2	R_2	Chi^2
low alloyed Ti	87	1.86×10^{-5}	13.6×10^{-6}	0.941	310	6.887×10^{-6}	31.7×10^{-6}	0.749	1.38×10^{7}	0.0007
Laser 65_3	29	2.38×10^{-5}	12.6×10^{-6}	0.904	1043	3.028×10^{-6}	4.76×10^{-6}	0.892	1.38×10^{7}	0.003
Laser 100_6.5	13	1.31×10^{-5}	13.1×10^{-6}	1	541	1.108×10^{-5}	39.4×10^{-6}	0.754	4.4×10^{6}	0.0008
Laser 100_4	30	2.09×10^{-5}	15.2×10^{-6}	0.929	701	1.97×10^{-6}	65.0×10^{-6}	0.5	1.67×10^{7}	0.002
laser 100_3	15	1.02×10^{-6}	10.2×10^{-6}	1	552	1.24×10^{-5}	63.9×10^{-6}	0.779	1.31×10^{7}	0.00057

The parameter R_e has a value from 15 to 87 $\Omega{\cdot}cm^2$ and is ascribed to electrolyte resistance. R_1 values are smaller than R_2 values as large values for R_2 are observed for tested materials, showing that the oxide film on Ti alloy and laser treated surfaces has a large resistance. It is seen that R_1 is less than 1 $k\Omega{\cdot}cm^2$. These values show that the porous layer formed has a very low resistance. CPE, denoted as Q_1 and Q_2 were recalculated using equation $C_1 = [R_1{}^{1-n} \times Q_1]^{1/n}$ [37] in order to compare capacitance values for Ti alloy. C_1 is attributed to the properties of outer oxide layer, while C_2 is in general increasing with the power of laser treatment and could be attributed to more compact inner dense layer.

In order to observe the long-term stability of films on Ti surfaces in corrosive solutions, electrochemical impedance measurements on low alloyed Ti and surface treated Ti were conducted after 24 h and subsequently every week in 3 week exposure time. The impedance magnitude at the measured low frequency limit, $|Z|_{f\to0}$ (f < 0.001 Hz) is presented in Figure 5. It is in the order of $M\Omega{\cdot}cm^2$ at initial time and steadily increases in the case of laser treated surfaces, or decreases in the case of low alloyed Ti surface and Ti sample treated with low power and high production speed (Ti_laser_65_3). The total impedance of the low alloyed Ti and laser treated Ti surfaces at different immersion times during long term experiment is presented in Figure 5.

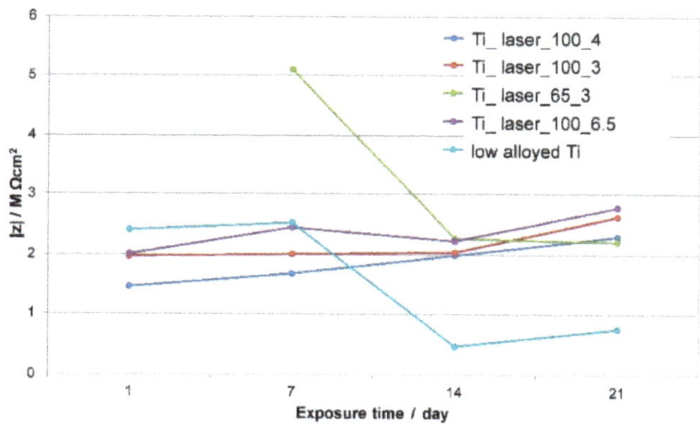

Figure 5. Total impedance |Z| of Ti surfaces in 0.1 M NaCl during 3-week exposure.

It can be observed that the total impedance of low alloyed Ti decreases with exposure time. The laser marked surfaces with the smallest power (Ti-laser-65-3) also experienced the total impedance decrease over exposure time. For all other observed laser treated surfaces, it can be observed that total impedance increases with time, pointing at the fact that the stable passive film was formed, which with exposure to aggressive electrolyte increases the corrosion stability of the laser treated films.

3.3. XPS Analysis

XPS analyses were performed on three samples: Non-treated low alloyed Ti sample, non-treated low alloyed Ti sample after exposure to 0.1 M NaCl solution for 3 weeks and laser marked Ti surface with 100% power and 3% production speed of a laser (sample Ti_laser_100_3), subsequently exposed to 0.1 M NaCl solution for 3 weeks. The surface of the last sample had a blue colour, as shown in Figure 1e.

XPS survey spectra were acquired on the surface of these samples and at the depth of 25 nm. XPS survey spectra are presented in Figure 6.

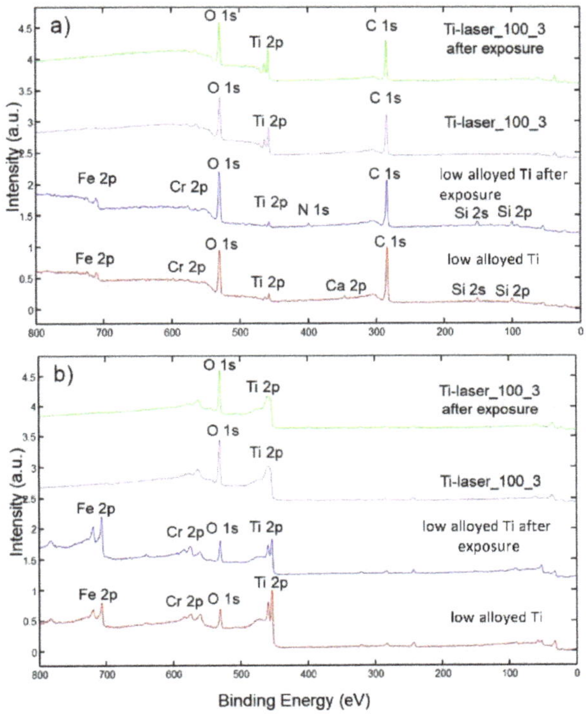

Figure 6. XPS survey spectra of low alloyed Ti before and after exposure to 0.1 M NaCl solution and laser marked Ti surface Ti-100-3 before and after exposure to 0.1 M NaCl solution; (**a**) at the surface and (**b**) at a depth of 25 nm.

Survey spectra for low alloyed Ti and that after exposure to the corrosive environment were similar (Figure 6a). No obvious difference in the survey spectra can be observed. Signals of O 1*s*, Ti 2*p* and Cr 2*p*, Fe 2*p*, Si 2*p*, Si 2*s*, as well as C 1*s* and Ca 2*p*, were present on the surface of low alloyed Ti (Figure 6a). This shows that surface treatment of low alloyed Ti sample yields Fe and Cr elements on the surface. At the surface of laser marked surface (Ti-laser_100_3) and exposed to corrosive environment only O, Ti and C were present with minor traces of N (Figure 6a).

In the subsurface region, at the depth of ~25 nm, Ti and Cr/Fe were present on low alloyed Ti samples, on laser marked surface only Ti and O were present (Figure 6b).

Figure 7 shows high energy resolution Ti 2*p* spectra of low alloyed Ti before exposure to 0.1 M NaCl solution and of laser marked Ti surface Ti-100-3 before exposure to 0.1 M NaCl solution at the surface and at a depth of 25 nm. The Ti $2p^{3/2}$ peak on the surface of both samples is at 458–459 eV what is assigned to Ti(4+) oxidation state in the surface TiO_2 oxide. After sputtering at depth of 25 nm on the laser non-marked sample Ti $2p^{3/2}$ peak was at 454 eV (Figure 7a) showing metallic phase of Ti. On the laser marked sample at depth of 25 nm Ti $2p^{3/2}$ peak showed many subpeaks ranging from 454 to 459 eV (Figure 7b) presenting different Ti-oxidation states from Ti(0) to Ti(4+). Such mixed spectrum of Ti oxidation states is a typical result of ion sputtering of TiO_2 when the reduction of Ti(4+) state is a consequence of preferential sputtering of oxygen from TiO_2 oxide [38]. However, also in this case, TiO_2 was present at depth of 25 nm which was the maximum depth reached during depth profiling.

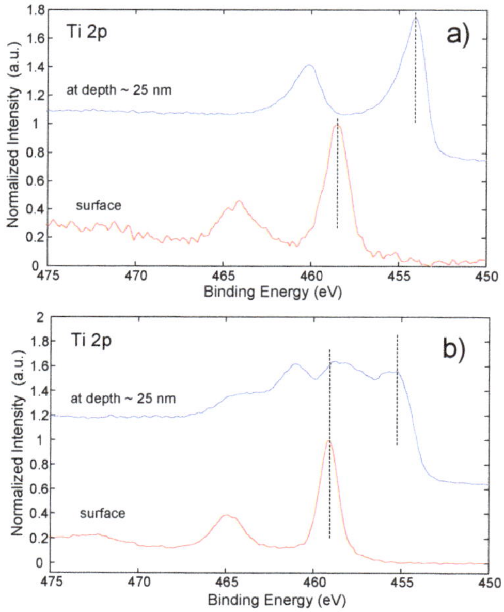

Figure 7. High energy resolution Ti 2*p* spectra of (**a**) low alloyed Ti before exposure to 0.1 M NaCl solution and (**b**) laser marked Ti surface Ti-100-3 before exposure to 0.1 M NaCl solution at the surface and at a depth of 25 nm.

Figure 8 shows the XPS depth profiles of four analysed samples presenting changes in chemical composition with a depth of up to ~25 nm. In Figure 8a,b, no curves for Cr are presented (~10 at %) although Cr was present in the surface layer, as proved by analyses after XPS depth profiling (Figure 6b). From Figure 6a, it follows that low alloyed Ti sample was covered by a ~10 nm thick mixed TiO_2/FeCr oxide layer (Figure 7a). On the surface, some traces of Ca and Si were also found. Beneath the surface oxide layer, a Ti–(Fe/Cr) layer was present. In the subsurface region oxygen and carbon concentrations did not decrease to 0% but they persist at significantly high level. We explained the presence of O and C as adsorption of O- and C-based species from residual vacuum atmosphere to fresh and very reactive Ti surface which was exposed during Ar bombardment. In addition, the relatively rough sample surface produced by shot peening induced shadowing effects for Ar ion bombardment resulting in not complete and non-uniform removal of adsorbed layer of O and C species. The adsorption effect of C and O was observed for all four samples being larger for laser non-treated samples (more metallic composition) than for laser treated samples containing less reactive surface oxide layer (described below).

The low alloyed Ti sample after exposure to the corrosive environment of 0.1 M NaCl had a similar surface composition as the surface low alloyed Ti before exposure. Low alloyed Ti after exposure was also covered by a ~10 nm thick mixed TiO_2/FeCr-oxide layer. On the surface, some N and Si were also present. The origin of C in the subsurface region is the same as described above. Beneath the surface oxide layer, a Ti–(Fe/Cr) layer was present. The concentration of Fe/Cr steel phase in the subsurface region was higher than on the low alloyed Ti before exposure, which may be related to the exposure with the 0.1 M NaCl solution.

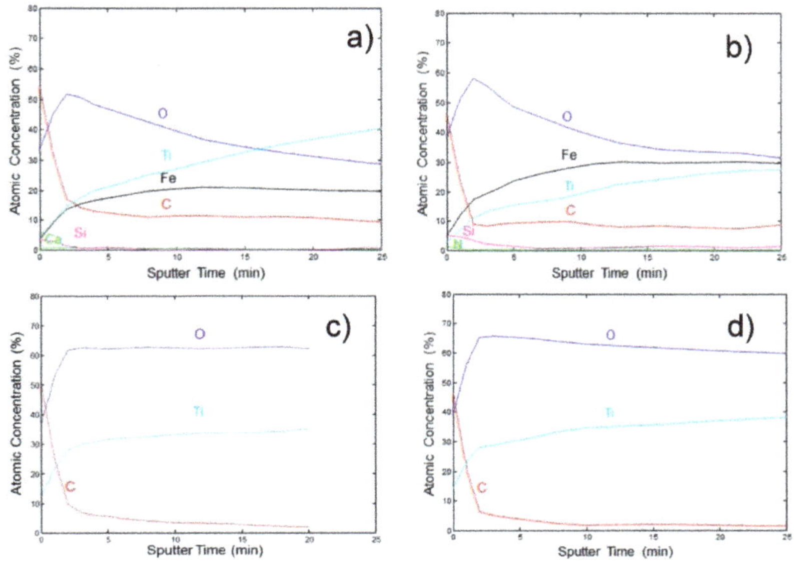

Figure 8. XPS depth profiles for (**a**) low alloyed Ti surface, (**b**) low alloyed Ti surface after exposure to corrosive environment, (**c**) laser marked Ti- Laser_100_3 sample before and (**d**) laser marked Ti-Laser_100_3 sample after exposure to corrosive environment. Sputtering rate was ~1 nm/min.

The surface and subsurface region of the laser marked sample (Ti-Laser_100_3) after exposure to the corrosive environment were substantially different from the samples of low alloyed Ti before and after exposure, which was related to different treatments. The laser marked surface of a sample was covered by a thicker TiO_2 layer when compared to low alloyed Ti samples. This thicker TiO_2 layer was related to the laser treatment. A notable difference in the laser treated sample with respect to low alloyed Ti samples was that no Fe and Cr, as well as no Si, were found on the TiO_2 surface layer. After exposure to 0.1 M NaCl, the laser treated surface was very similar to the surface before exposure to the corrosive environment.

XPS analysis showed that low alloyed Ti samples were different from the laser marked Ti surface. The main difference was in the chemical structure of the surface oxide film. This was a 10 nm thick layer of Ti–(Fe/Cr) oxides for the low alloyed Ti surface. This oxide layer did not change much with exposure to the aggressive 0.1 M NaCl solution. The TiO_2 layer on the laser treated sample was thicker (at least 25 nm what was the largest depth reached by depth profiling) and did not contain elements like Fe, Cr and Si introduced by surface modification of the low alloyed Ti sample.

The more expressed corrosion resistance evidenced from Figure 5 for the laser treated Ti surface over the non-treated surface can be explained by the presence of Fe and Cr over Ti in the oxide film present on the non-treated surface of low alloyed Ti sample.

Comparing the results of XPS analysis, it can be concluded that mainly TiO_2 film of thicknesses more than 25 nm improves corrosion properties of laser treated surfaces, since TiO_2 film, of protective nature, is formed by laser treatment.

4. Conclusions

Microstructural, electrochemical and surface analyses were performed on laser marked low alloyed Ti surfaces in order to study the effect of laser treatment on the microstructural, physical and corrosion properties of such surfaces in a corrosive environment.

- The depth of microstructural changes in the cross-sections of laser treated surfaces was related to the laser power and production speed of the laser. It was the highest for 100% power and the lowest production speed of the laser.
- Corrosion properties were studied by potentiodynamic measurements and electrochemical impedance spectroscopy. It was found that laser treated surfaces exhibited lower current densities in the passive region, while the stability of such surfaces is enhanced in long-term exposure in an aggressive environment.
- XPS analysis showed that the low alloyed Ti surface was different from the laser marked surface.
- A mixture of Ti and Fe/Cr oxides of thickness of 10 nm was found on low alloyed Ti surfaces, however mainly pure TiO_2 oxide layer was detected on laser marked surfaces. TiO_2 oxide was thicker (25 nm) than those on laser non-treated surface (about 10 nm).
- Different constitution and thickness of oxide layer of low alloyed Ti surface (Cr and Fe presence in TiO_2 surface film) and laser marked surfaces (thick TiO_2) affected long-term corrosion susceptibility. Low alloyed Ti surfaces were less stable than laser marked surfaces.

Author Contributions: Conceptualization, T.K. and D.K.; Investigation, J.K., D.K. and T.K.; Writing—Original Draft Preparation, T.K. and D.K.; Supervision, A.L.; Writing—Review and Editing, T.K., D.K., A.L. and J.K.

Funding: This research was funded by the Slovenian Research Agency (research core Funding No. P2-0270, P2-0082 and P2-0273, and Project No. L2-8181).

Acknowledgments: The help of the laboratory staff, Viljem Kuhar and Rok Tomšič, is hereby gratefully acknowledged.

Conflicts of Interest: The authors declare no conflict of interest.

References

1. Ng, T.; Yeo, S. Aesthetic laser marking assessment using luminance ratios. *Opt. Lasers Eng.* **2001**, *35*, 177–186. [CrossRef]
2. Zheng, H.; Lim, G.C. Laser-effected darkening in TPEs with TiO_2 additives. *Opt. Lasers Eng.* **2004**, *41*, 791–800. [CrossRef]
3. Liu, C.; Lu, Y.; Xiong, Y.; Zhang, Q.; Cao, Z. Recognition of laser-marked quick response codes on polypropylene surfaces. *Polym. Degrad. Stab.* **2018**, *147*, 115–122. [CrossRef]
4. Zelenska, K.; Zelensky, S.E.; Poperenko, L.V.; Kanev, K.; Mizeikis, V.; Gnatyuk, V.A. Thermal mechanisms of laser marking in transparent polymers with light-absorbing microparticles. *Opt. Laser Technol.* **2016**, *76*, 96–100. [CrossRef]
5. Fujimoto, S.; Tsujino, K.; Shibata, T. Growth and properties of Cr-rich thick and porous oxide films on Type 304 stainless steel formed by square wave potential pulse polarisation. *Electrochim. Acta* **2001**, *47*, 543–551. [CrossRef]
6. Antonczak, A.J.; Stepak, B.; Koziol, P.E.; Abramski, K.M. The influence of process parameters on the laser-induced coloring of titanium. *Appl. Phys. A* **2014**, *115*, 1003–1013. [CrossRef]
7. Del Pino, A.P.; Serra, P.; Morenza, J. Oxidation of titanium through Nd: YAG laser irradiation. *Appl. Surf. Sci.* **2002**, *197*, 887–890. [CrossRef]
8. Diamanti, M.V.; Del Curto, B.; Pedeferri, M. Interference colors of thin oxide layers on titanium. *Color Res. Appl.* **2018**, *33*, 221–228. [CrossRef]
9. Veiko, V.; Odintsova, G.; Gorbunova, E.; Ageev, E.; Shimko, A.; Karlagina, Y.; Andreeva, Y. Development of complete color palette based on spectrophotometric measurements of steel oxidation results for enhancement of color laser marking technology. *Mater. Des.* **2016**, *89*, 684–688. [CrossRef]
10. Penide, J.; Quintero, F.; Riveiroa, A.; Fernández, A.; del Val, J.; Comesaña, R.; Lusquiños, F.; Pou, J. High contrast laser marking of alumina. *Appl. Surf. Sci.* **2015**, *336*, 118–128. [CrossRef]
11. Durdu, S.; Usta, M.; Berkem, A.S. Bioactive coatings on Ti6Al4V alloy formed by plasma electrolytic oxidation. *Surf. Coat. Technol.* **2016**, *301*, 85–93. [CrossRef]
12. O'Hana, S.; Pinkerton, A.J.; Shoba, K.; Gale, A.W.; Li, L. Laser surface colouring of titanium for contemporary jewellery. *Surf. Eng.* **2008**, *24*, 147–153. [CrossRef]

13. Akman, E.; Cerkezoglu, E. Compositional and micro-scratch analyses of laser induced colored surface of titanium. *Opt. Lasers Eng.* **2016**, *84*, 37–43. [CrossRef]
14. Amara, E.; Haid, F.; Noukaz, A. Experimental investigations on fiber laser color marking of steels. *Appl. Surf. Sci.* **2015**, *351*, 1–12. [CrossRef]
15. Antończak, A.J.; Kocoń, D.; Nowak, M.; Kozioł, P.; Abramski, K.M. Laser-induced colour marking—Sensitivity scaling for a stainless steel. *Appl. Surf. Sci.* **2013**, *264*, 229–236. [CrossRef]
16. Antończak, A.J.; Skowroński, Ł.; Trzcinski, M.; Kinzhybalo, V.V.; Łazarek, Ł.K.; Abramski, K.M. Laser-induced oxidation of titanium substrate: Analysis of the physicochemical structure of the surface and sub-surface layers. *Appl. Surf. Sci.* **2015**, *325*, 217–226. [CrossRef]
17. Astarita, A.; Genna, S.; Leone, C.; Minutolo, F.M.C.; Squillace, A.; Velotti, C. Study of the laser marking process of cold sprayed titanium coatings on aluminium substrates. *Opt. Laser Technol.* **2016**, *83*, 168–176. [CrossRef]
18. Qi, J.; Wang, K.; Zhu, Y. A study on the laser marking process of stainless steel. *J. Mat. Process. Technol.* **2003**, *139*, 273–276. [CrossRef]
19. Salou, L.; Hoornaert, A.; Louarn, G.; Layrolle, P. Enhanced osseointegration of titanium implants with nanostructured surfaces: An experimental study in rabbits. *Acta Biomater.* **2015**, *11*, 494–502. [CrossRef]
20. Veiko, V.; Karlagina, Y.; Moskvin, M.; Mikhailovskii, V.; Odintsova, G.; Olshin, P.; Pankin, D.; Romanov, V.; Yatsuk, R. Metal surface coloration by oxide periodic structures formed with nanosecond laser pulses. *Opt. Lasers Eng.* **2017**, *96*, 63–67. [CrossRef]
21. Teng, K.; Delplancke, J.-L.; Zhang, J.; O'Keefe, T.J. Electrochemical characterization of copper deposited on plasma and thermally modified titanium surfaces. *Metall. Mater. Trans. B* **1998**, *29*, 749–754. [CrossRef]
22. Wang, H.-Y.; Zhu, R.-F.; Lu, Y.-P.; Xiao, G.-Y.; Ma, X.-N.; Li, Y. Structures and properties of layered bioceramic coatings on pure titanium using a hybrid technique of sandblasting and micro-arc oxidation. *Appl. Surf. Sci.* **2013**, *282*, 271–280. [CrossRef]
23. Zhao, B.; Zhao, B.H.; Zhang, W.; Wang, D.N.; Feng, W.; Liu, Y.; Lin, Z.; Du, K.Q.; Deng, C.F. Effect of Zn content on cytoactivity and bacteriostasis of micro-arc oxidation coatings on pure titanium. *Surf. Coat. Technol.* **2013**, *228*, S428–S432. [CrossRef]
24. Diamanti, M.V.; Del Curto, B.; Masconale, V.; Pedeferri, M. Production and anodic colouring of newly-designed titanium jewels. *Colour Des. Creat.* **2010**, *16*, 1–9.
25. Lavisse, L.; Grevey, D.; Langlade, C.; Vannes, B. The early stage of the laser-induced oxidation of titanium substrates. *Appl. Surf. Sci.* **2002**, *186*, 150–155. [CrossRef]
26. Banerjee, D.; Williams, J. Perspectives on titanium science and technology. *Acta Mater.* **2013**, *61*, 844–879. [CrossRef]
27. Li, Z.; Zheng, H.Y.; Teh, K.M.; Liu, Y.C.; Lim, G.C.; Seng, H.L.; Yakovlev, N.L. Analysis of oxide formation induced by UV laser coloration of stainless steel. *Appl. Surf. Sci.* **2009**, *256*, 1582–1588. [CrossRef]
28. Kasman, S. Impact of parameters on the process response: A Taguchi orthogonal analysis for laser engraving. *Measurement* **2013**, *46*, 2577–2584. [CrossRef]
29. Adams, D.; Murphy, R.D.; Saiz, D.J.; Hirschfeld, D.A.; Rodriguez, M.A.; Kotula, P.G.; Jared, B.H. Nanosecond pulsed laser irradiation of titanium: Oxide growth and effects on underlying metal. *Surf. Coat. Technol.* **2014**, *248*, 38–45. [CrossRef]
30. Kung, K.-C.; Lee, T.-M.; Lui, T.-S. Bioactivity and corrosion properties of novel coatings containing strontium by micro-arc oxidation. *J. Alloys Compd.* **2010**, *508*, 384–390. [CrossRef]
31. Dong, H.; Li, X.Y. Oxygen boost diffusion for the deep-case hardening of titanium alloys. *Mat. Sci. Eng. A* **2000**, *280*, 303–310. [CrossRef]
32. Valete, S.; Steyer, P.; Richard, L.; Forest, B.; Donnet, C.; Audoarrd, E. Influence of femtosecond laser marking on the corrosion resistance of stainless steels. *Appl. Surf. Sci.* **2006**, *25*, 4696–4701. [CrossRef]
33. Calvert, K.; Kosaka, Y. Evaluation of titanium alloys after High temperature air exposure. In Proceedings of the 13th World Conference on Titanium, San Diego, CA, USA, 16–20 August 2015; The Minerals, Metals & Materials Society: Pittsburgh, PA, USA, 2016.
34. Tait, W.S. *An Introduction to Electrochemical Corrosion Testing for Practicing Engineers and Scientists*; Ohio State University: Columbus, OH, USA, 1994.
35. Pan, J.; Thierry, D.; Leygraf, C. Electrochemical impedance spectroscopy study of passive oxide film on titanium for implant application. *Electrochim. Acta* **1996**, *41*, 1143–1153. [CrossRef]

36. Raistrick, I.D.; MacDonald, J.R.; Francschetti, D.R. *Impedance Spectroscopy Emphasizing Solid Materials and Systems*; MacDonald, J.R., Ed.; John Wiley & Sons: New York, NY, USA, 1987.
37. Kek-Merl, D.; Lappalainen, J.; Tuller, H.L. Electrical properties of nanocrystalline CeO_2 thin films deposited by in situ pulsed laser deposition. *J. Electrochem. Soc.* **2006**, *153*, J15–J20. [CrossRef]
38. Hashimonto, S.; Tanaka, A. Alteration of Ti 2*p* XPS spectrum for titanium oxide by low-energy Ar ion bombardment. *Surf. Interface Anal.* **2002**, *34*, 262–265. [CrossRef]

coatings

MDPI

Article

The Corrosion Inhibition of AA6082 Aluminium Alloy by Certain Azoles in Chloride Solution: Electrochemistry and Surface Analysis

Klodian Xhanari [1,2] and Matjaž Finšgar [1,*]

1 Faculty of Chemistry and Chemical Engineering, University of Maribor, Smetanova ulica 17, Maribor 2000, Slovenia; klodian.xhanari@fshn.edu.al
2 Faculty of Natural Sciences, University of Tirana, Boulevard "Zogu I", Tirana 1001, Albania
* Correspondence: matjaz.finsgar@um.si

Received: 15 May 2019; Accepted: 10 June 2019; Published: 13 June 2019

Abstract: The corrosion inhibition effect of five azole compounds on the corrosion of an AA6082 aluminium alloy in 5 wt.% NaCl solution at 25 and 50 °C was investigated using weight loss and electrochemical measurements. Only 2-mercaptobenzothiazole (MBT) showed a corrosion inhibition effect at both temperatures and was further studied in detail, including with the addition of potassium iodide as a possible intensifier. Surface analysis of the MBT surface layer was performed by means of attenuated total reflectance Fourier transform infrared spectroscopy, X-ray photoelectron spectroscopy, and time-of-flight secondary ion mass spectrometry techniques. The hydrophobicity of the MBT surface layer was also investigated.

Keywords: 2-mercaptobenzothiazole; aluminium alloy; AA6082; corrosion inhibitor; chloride solution; automotive industry

1. Introduction

The AA6xxx series aluminium alloys (also known as Al–Mg–Si alloys) are heat-treatable wrought alloys, which, in addition to their high thermal and electrical conductivity and low specific weight, possess moderately high strength [1,2]. These properties, combined with their good formability and high resistance to general corrosion, make these alloys very suitable for industrial use, especially in the automotive and aerospace industries [3,4]. However, intergranular corrosion is one of the most reported forms of corrosion for AA6xxx series aluminium alloys [5–11]. The authors reported a close connection between the amount of Cu present in the alloys and their susceptibility to intergranular corrosion. The Cu content is also related to the pitting corrosion susceptibility of these alloys in chloride-containing solutions [3]. Due to the fact that the AA6082 aluminium alloy is used to fabricate hot extruded automotive parts [3,4], the corrosion resistance of this alloy in chloride-containing solutions has been reported in several studies. Cicolin et al. [12] reported that an increase in chloride concentration (from 0.01 to 1.00 M NaCl) made the AA6082 aluminium alloy more susceptible to pitting than to intergranular corrosion. The same behaviour was also observed when increasing the solution's pH or the amount of dissolved oxygen. The effect of pH on the stress corrosion cracking of the AA6082 aluminium alloy in a 0.3 M NaCl solution was also studied by Panagopoulos et al. [13]. The authors reported higher corrosion susceptibility in basic and acid solutions, while a higher resistance was observed for neutral solutions. Trdan and Grum [14] used electrochemical measurements to study the influence of laser shock peening (LSP) treatment on the corrosion resistance of AA6082-T651 aluminium alloy immersed in a 0.6 M NaCl solution. The passivity region of the AA6082-T651 aluminium alloy samples increased after LSP treatment. Moreover, the LSP-treated samples showed a repassivation ability.

Recently, we reported on the influence of chloride concentration (1 and 5 wt.% NaCl solution) and temperature (25 and 50 °C) on the corrosion behaviour of the AA6082 aluminium alloy in chloride-containing solutions [15]. A corrosion rate increase was observed when increasing the chloride concentration and temperature. Both general and pitting corrosion were more pronounced at the highest temperature (50 °C). Moreover, the corrosion of the AA6082 aluminium alloy samples under the above-mentioned conditions followed kinetic-controlled processes [15].

Surface modification by either laser shock peening without ablative coating [14] or polyvinyl alcohol fibrous coating [16], as well as by the application of a titania coating [17], was employed to mitigate the corrosion of AA6082 aluminium alloys in chloride-containing solutions. In addition, sodium molybdate dehydrate was used by Panagopoulos et al. [18] as a corrosion inhibitor for the AA6082 aluminium alloy in a 0.01 M NaCl solution. Although a large number of organic compounds and natural products have been employed as corrosion inhibitors for aluminium alloys in acid, basic, and chloride-containing solutions [19–21], the literature is scarce regarding the use of organic compounds as corrosion inhibitors for the AA6082 aluminium alloy. Azole compounds, mainly derivatives of imidazole and benzotriazole, have shown good corrosion inhibition effects in chloride-containing solutions, mostly for copper [22–31] and, in a few cases, for steel [32–35]. Moreover, several azole compounds have been used as corrosion inhibitors, mostly for pure aluminium and the AA2024 aluminium alloy in chloride-containing solutions [36–40]. Hitherto, to the best of our knowledge, there have been no reports in the literature on the use of azole compounds as corrosion inhibitors for the AA6082 aluminium alloy in chloride-containing solutions.

In this work, electrochemical impedance spectroscopy (EIS) measurements were first used to evaluate the effectiveness of five azole compounds, i.e., 2-mercaptobenzothiazole (MBT), 2-mercaptobenzoxazole (MBO), 2-mercaptobenzimidazole (MBI), benzotriazole (BTA), and 3-amino-1H-1,2,4-triazole (3-AT), as corrosion inhibitors for the AA6082 aluminium alloy during 10 h of immersion in 5 wt.% NaCl solution at 50 °C, with and without the addition of KI as a possible intensifier. The corrosion inhibition effectiveness of the above-mentioned compounds was tested at 50 °C, a common temperature employed in the automotive industry. According to the international standard ISO 9227 [41], the neutral salt spray tests should be performed at 35 ± 2 °C. However, the same standard states that the cooper-accelerated acetic salt spray tests should be performed at 50 ± 2 °C. In this work, the inhibition effect of the studied azole compounds in the corrosion of AA6082 aluminium alloy samples in 5 wt.% NaCl solution was first studied at 50 °C to simulate an even more corrosive environment and test the worst-case scenario. Next, only the compounds found to show a corrosion inhibition effect at 50 °C were tested at 25 °C [41], with and without the addition of KI, using electrochemical techniques (including open circuit potential, potentiodynamic curve, and EIS measurements) and the weight loss method. The adsorption of the effective azole compound was confirmed by attenuated total reflectance Fourier transform infrared spectroscopy (ATR-FTIR), X-ray photoelectron spectroscopy (XPS), and time-of-flight secondary ion mass spectrometry (ToF-SIMS) measurements. Finally, contact angle measurements were performed on AA6082 aluminium alloy samples immersed in 5 wt.% NaCl solution at 25 °C.

2. Materials and Methods

2.1. Sample Preparation

The AA6082 aluminium alloy with the composition given in Table 1 was supplied by Rocholl GmbH, Aglasterhausen, Germany. Rectangle-shaped samples (50 by 20 by 1 mm) were used for the weight loss measurements, while disc-shaped samples (diameter 15 mm) were used for the electrochemical and surface characterization measurements [15,22,28,33].

Table 1. Chemical composition of AA6082 aluminium alloy as specified by the supplier.

Element	Si	Fe	Cu	Mn	Mg	Cr	Zn	Ti	Al
Content (wt.%)	0.90	0.35	0.07	0.45	0.90	0.04	0.05	0.02	balance

Prior to performing weight loss tests, the samples were first cleaned ultrasonically in a 50 vol.% acetone/50 vol.% ultrapure water bath and then dried under a stream of air. The samples used for the electrochemical measurements were first ground with different SiC papers (320, 500, 800, 1000, 2400, and 4000 grit), supplied by Struers (Ballerup, Denmark), and then rinsed with ultrapure water (resistivity of 18.2 MΩ cm at 25 °C) obtained by means of the Milli-Q system (Millipore Corporation, Burlington, MA, USA). Finally, the samples were ultrasonically cleaned in a 50 vol.% acetone/50 vol.% ultrapure water bath and then dried under a stream of air.

2.2. Solution Preparation

All the solutions used in this study were prepared using ultrapure water. The 97 wt.% pure 2-mercaptobenzothiazole (MBT), 95 wt.% pure 2-mercaptobenzoxazole (MBO), 98 wt.% pure 2-mercaptobenzimidazole (MBI), and 99 wt.% pure benzotriazole (BTA) were supplied by Sigma-Aldrich (St. Louis, MO, USA), while the 95 wt.% pure 3-amino-1H-1,2,4-triazole (3-AT) was supplied by ACROS Organics (Morris County, NJ, USA). The chemical structures of these compounds are given in Figure 1.

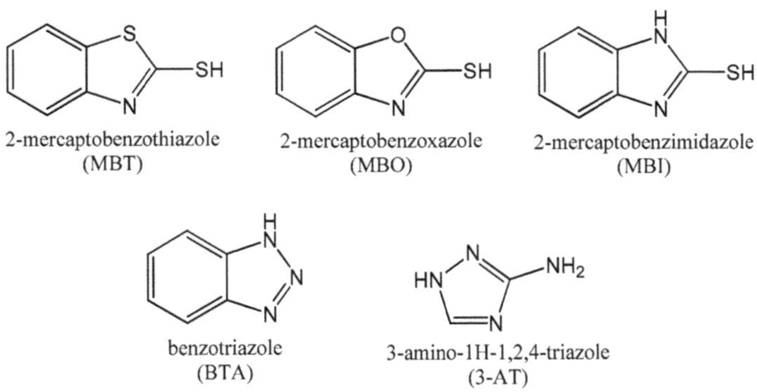

Figure 1. Structures of the five azole compounds used in this study.

Acetone, NaCl, and KI (for analysis, ACS quality) were supplied by Carlo Erba Reagents (Milan, Italy). The AA6082 aluminium alloy samples were immersed in 5 wt.% NaCl solution containing a) 0.3 mM MBT, b) 1.0 mM MBO, c) 1.0 mM MBI, d) 10.0 mM 3-AT, e) 1.0 mM BTA, and f) 10.0 mM BTA. The choice of the concentrations for the MBT, MBO, and MBI is related to their solubility limit in 5 wt.% NaCl solution. A concentration of 1 mM for these compounds is close to the 0.1 wt.% concentration that is commonly used for corrosion tests in industry [42].

2.3. Weight Loss Measurements

The AA6082 aluminium alloy samples were weighed before immersion in the tested solutions. After seven days of immersion, the samples were first rinsed with ultrapure water, and then a fibre-bristle brush was used to remove the corrosion products from their surface. Before weighing the samples again, they were rinsed once more with ultrapure water and dried under a stream of air. Dixon's and Grubb's statistical tests [43] were used to remove any outliers, and an average value of at least six replicate measurements was reported.

2.4. Electrochemical Measurements

The AA6082 aluminium alloy samples immersed in 5 wt.% NaCl solution with each of the azole compounds were subjected to electrochemical measurements under stagnant conditions at 25 and 50 °C. The working electrode consisted of the AA6082 aluminium alloy sample (prepared as described above) embedded in a Teflon holder (PAR, Cambridge, UK) with a 1 cm^2 area exposed to the solution [15,22,28,33]. A saturated calomel electrode, SCE (0.244 V vs. SHE, the standard hydrogen electrode), and a graphite rod were used as the reference and the counter electrode, respectively. A Gamry 600™ potentiostat/galvanostat controlled by a Gamry Framework electrochemical program (Gamry Instruments, Warminster, PA, USA) was used to perform all the electrochemical measurements. The results obtained were analysed using the Gamry EChem Analyst 6.30 software.

The EIS spectra were recorded at the open circuit potential (E_{oc}), in the frequency range from 1 MHz to 10 mHz, using a signal with a 10 mV (peak to peak) amplitude and 10 points/decade [15,22,28,33]. The EIS measurements were performed in sequence after 1, 3, 5, 7, and 10 h of immersion in 5 wt.% NaCl solution with and without different azole compounds. The E_{oc} was determined from the time of immersion in sequence before each of the EIS measurements and just before the potentiodynamic curve measurements. The potentiodynamic curves were recorded after 11 h of immersion, starting from −0.250 V vs. E_{oc} and continuing in the anodic direction with a potential scan rate of 0.1 mV s^{-1}. At least three replicate measurements were performed for all the electrochemical techniques and a representative curve is given in each case.

2.5. Surface Analysis

The samples prepared with the same procedure as the electrochemical measurements were subjected to surface analysis after immersion in 5 wt.% NaCl solution with and without the presence of the azole compounds. The ATR-FTIR, XPS, ToF-SIMS, and contact angle measurements were used to characterize the surface of the AA6082 aluminium alloy samples after immersion. The ATR-FTIR and contact angle measurements were performed after 31 days of immersion, while for the XPS (PHI 5600, Physical Electronics, Inc., Chanhassen, MN, USA) and ToF-SIMS (ToF-SIMS 5, ION-TOF, Münster, Germany) measurements, a 1 h immersion time was used. A Shimadzu IRAffinity-1 (Kyoto, Japan) spectrometer was used for the ATR-FTIR measurements. The instrumental setup for the XPS and ToF-SIMS measurements were the same as reported previously [44]. The hydrophobicity of the AA6082 aluminium alloy samples after immersion in 5 wt.% NaCl solutions containing the most effective azole compound (higher corrosion inhibition effectiveness) with and without the addition of 0.1 wt.% KI was tested using an OCA 35 Dataphysic contact angle analyser (Filderstadt, Germany), with the respective software (SCA 20). The average value of at least three replicate contact angle measurements obtained at different spots of the sample's surface was reported (outliers were checked with Dixon's and Grubb's statistical tests, and any present outliers were removed from the calculation of the average value [43]).

3. Results and Discussion

3.1. Electrochemical Measurements

The AA6082 aluminium alloy samples immersed in 5 wt.% NaCl solution containing the MBI, MBO, MBT, 3-AT, and BTA azole compounds, as described above, were first tested by means of EIS measurements at 50 °C to simulate the worst-case corrosion scenario and to select the best candidates for lower temperature studies. The Nyquist plots of the AA6082 aluminium alloy samples immersed in 5 wt.% NaCl solution with different amounts of the studied azole compounds (Figure 2) obtained at 50 °C indicated that only MBT and 3-AT showed a corrosion inhibition effect. However, the parameters obtained from the EIS fitting procedure showed that the addition of MBT and 3-AT resulted in only a slight increase in the corrosion resistivity of the AA6082 aluminium alloy samples (the corrosion inhibition effectiveness calculated based on fitted R_p values as described below was found to be lower than 25%). For all the other azole compounds lower R_p values were obtained compared with the non-inhibited system indicating that these compounds are not corrosion inhibitors for the AA6082 aluminium alloy.

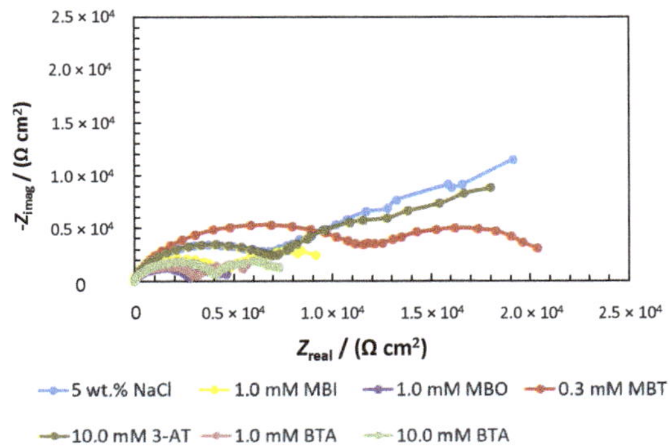

Figure 2. The Nyquist plots of the AA6082 aluminium alloy samples immersed at 50 °C in 5 wt.% NaCl solution containing different concentrations of five azole compounds.

Next, the corrosion inhibition effectiveness of MBT and 3-AT was further tested at 25 °C. No significant corrosion inhibition effect was found for 10.0 mM 3-AT in 5 wt.% NaCl solution at 25 °C, the main reason being the lower corrosion rate of the base aluminium alloy, which becomes similar to the corrosion rate of the alloy in the solution with the addition of 3-AT. However, MBT at 0.3 mM showed a corrosion inhibition effect at 25 °C. Due to the latter, MBT was electrochemically tested in detail, including open circuit potential, EIS, and potentiodynamic curve measurements, as presented below.

3.1.1. Open Circuit Potential Measurements

The open circuit potential (E_{oc}) of the AA6082 aluminium alloy samples immersed in 5 wt.% NaCl solution containing 0.3 mM MBT or a combination of 0.3 mM MBT and 0.1 wt.% KI was measured at 25 °C in sequence before each EIS measurements (starting from the moment of immersion) and just before the potentiodynamic curve was recorded. The last measured E_{oc} for each of the systems was considered to be the corrosion potential (E_{corr}). The open circuit potential values obtained for each system are presented in Figure 3 as a function of the immersion time.

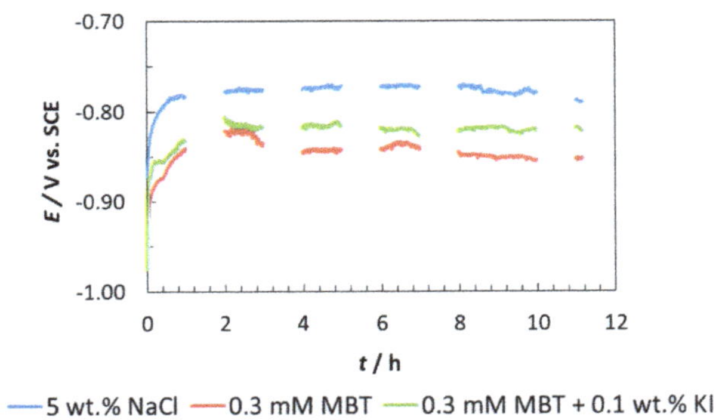

Figure 3. The influence of the immersion time on the open circuit potential values of the AA6082 aluminium alloy samples immersed in 5 wt.% NaCl solution with and without 0.3 mM MBT or a combination of 0.3 mM MBT and 0.1 wt.% KI at 25 °C. The discontinuous parts of the curves indicate the moments when the EIS measurements were performed.

Figure 3 shows that for the non-inhibited AA6082 aluminium alloy samples immersed in 5 wt.% NaCl solution the E_{oc} moved to more positive potentials from the moment of immersion until the first hour, and then no significant changes were observed (a steady state was achieved). For this system, E_{corr} was found to be −0.789 V. The addition of 0.3 mM MBT to 5 wt.% NaCl solution resulted in a shift of the E_{oc} to slightly more negative potentials compared with the non-inhibited system. The E_{oc} was first transferred to more positive potentials in the first hour of immersion and then slightly shifted to more negative potentials from the first to the third hour of immersion. For the inhibited system, a steady state was reached after 3 h of immersion. The E_{corr} for this system was found to be −0.852 V.

The steady state for the AA6082 aluminium alloy samples immersed in 5 wt.% NaCl solution containing 0.3 mM MBT with the addition of 0.1 wt.% KI was also achieved after 1 h of immersion. No significant shift was observed when comparing the open circuit potential curves for the inhibited system with and without the addition of 0.1 wt.% KI. This is reflected also in the E_{corr} value, which was −0.821 V for the inhibited system containing 0.1 wt.% KI.

The shift to more negative potentials when 0.3 mM MBT or a combination of 0.3 mM MBT and 0.1 wt.% KI were added to 5 wt.% NaCl solution indicates that MBT primarily inhibits the cathodic reaction of the corrosion couple, i.e., the reduction of the dissolved oxygen in the solution and the hydrogen evolution (which is most likely not intensive since for the 5 wt.% NaCl solution containing 0.3 mM MBT the pH was 5.07, while no significant change of the pH (pH = 5.11) was measured when 0.1 wt.% KI was added to the solution).

3.1.2. EIS Measurements

Figure 4 presents the EIS response of the AA6082 aluminium alloy samples after 1–10 h of immersion at 25 °C in 5 wt.% NaCl solution (Figure 4a–c), in 5 wt.% NaCl solution containing 0.3 mM MBT (Figure 4d–f), and in 5 wt.% NaCl solution containing 0.3 mM MBT with the addition of 0.1 wt.% KI (Figure 4g–i). The two distinctive patterns observed in the Bode modulus plot (Figure 4b,e) in the high and middle frequency regions and the respective behaviour of the systems associated with those patterns have previously been described in detail [15,22,28,33].

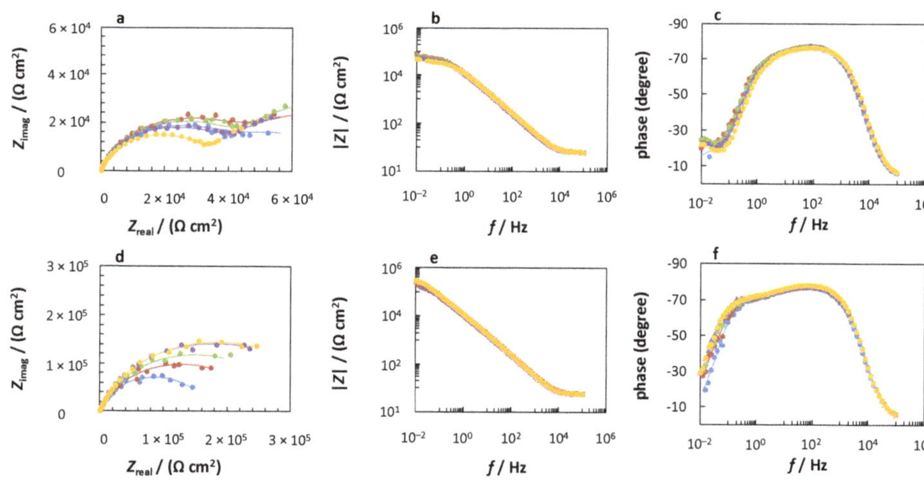

Figure 4. *Cont.*

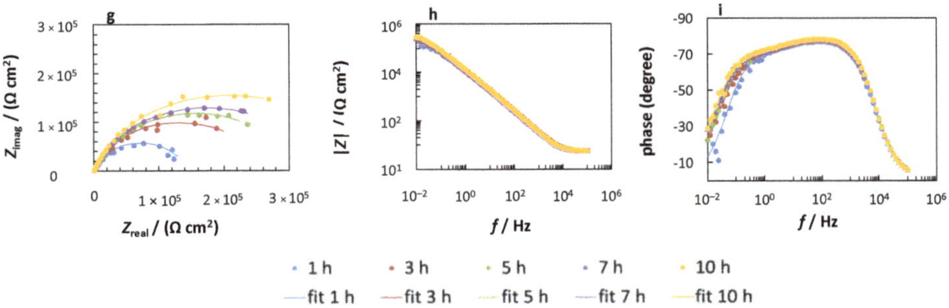

Figure 4. The measured electrochemical impedance spectroscopy (EIS) spectra (dotted) and the respective fitting (in continuous lines) for the AA6082 aluminium alloy samples immersed for 1–10 h, at 25 °C, in (**a–c**) 5 wt.% NaCl solution, (**d–f**) 5 wt.% NaCl solution containing 0.3 mM MBT, and (**g–i**) 5 wt.% NaCl solution containing 0.3 mM MBT with the addition of 0.1 wt.% KI.

Several equivalent electrical circuit (EEC) models were taken into consideration to fit the EIS spectra obtained. Figure 5 presents the nested $R_\Omega(Q_1(R_1(Q_2(R_2))))$ EEC model selected to fit the EIS response as the error of the fitting procedure (χ^2) using this EEC was the lowest among all EECs tested. The same EEC model was used previously to fit the EIS response of AA6082 aluminium alloy samples, under the same conditions, without the addition of azole compounds and intensifiers [15]. A combination of resistances and constant phase elements (represented by Q, which describes the non-ideal capacitance C) was used to describe the two relaxation processes in this EEC model. The uncompensated resistance (R_Ω) is largely influenced by the solution resistance. The processes occurring at the surface layer (a combination of the oxide layer and the adsorbed inhibitor) are described by the surface layer resistance (R_1) and capacitance (represented by Q_1). The charge-transfer resistance (R_2) and double-layer capacitance (represented by Q_2) describe the processes occurring at the metal/solution interface.

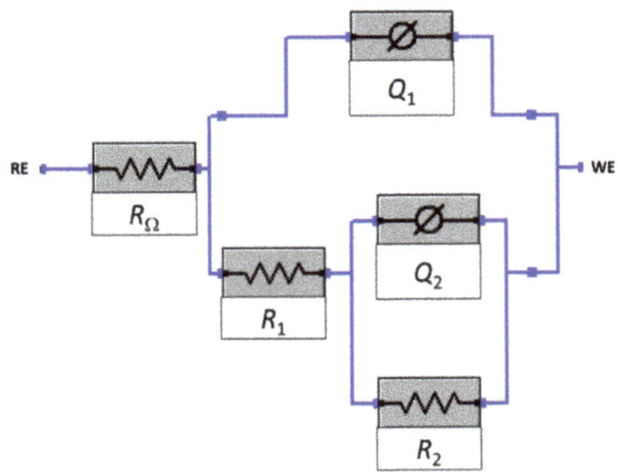

Figure 5. The $R_\Omega(Q_1(R_1(Q_2(R_2))))$ equivalent electrical circuit (EEC) model used to fit the EIS response of the AA6082 aluminium alloy samples immersed in 5 wt.% NaCl solution with and without 0.3 mM MBT or a combination of 0.3 mM MBT and 0.1 wt.% KI.

Table 2 shows the average values of the parameters obtained from the fitting procedure using the nested $R_\Omega(Q_1(R_1(Q_2(R_2))))$ EEC model presented in Figure 5. The capacitance values (C_x) were calculated from the respective non-ideal capacitance (Q_x) and resistance (R_x) values as $C_x = ((R_x\,Q_x)^{1/n_x})/R_x$. The thickness of the surface layer (d) is inversely proportional to the surface layer capacitance (C_1) [33].

The C_1 values of the AA6082 aluminium alloy samples immersed in 5 wt.% NaCl solution decreased with an increase in immersion time for both systems, i.e., with and without the addition of 0.1 wt.% KI (Table 2). This resulted in an increase in the surface layer thickness with increasing immersion time, as more of the inhibitor is adsorbed on the surface of the AA6082 aluminium alloy samples.

Table 2. The parameters obtained from fitting the EIS response of the AA6082 aluminium alloy samples immersed in 5 wt.% NaCl solution containing 0.3 mM MBT or a combination of 0.3 mM MBT and 0.1 wt.% KI, after 1–10 h of immersion, at 25 °C. In addition, the fitted parameters from the EIS spectra of the AA6082 aluminium samples in 5 wt.% NaCl solution, previously reported in [15], are given for comparison. Units: χ^2 [$\times10^{-3}$], R_Ω [Ω cm^2], R_1, R_2, and R_p [kΩ cm^2], Q_1 and Q_2 [$\mu\Omega^{-1}$ cm^{-2} s^n], and C_1 and C_2 [μF cm^{-2}].

Immersion Time [h]	χ^2	R_Ω	Q_1	n_1	R_1	C_1	Q_2	n_2	R_2	C_2	R_p
5 wt.% NaCl solution											
1	0.84	5.90	217.00	0.69	39.70	581.50	17.2	0.86	43.30	16.46	56.16
3	0.98	5.84	276.00	0.85	48.00	440.16	16.8	0.86	52.30	16.45	100.30
5	0.61	5.69	396.00	0.99	46.40	401.84	16.5	0.86	50.80	15.96	97.20
7	0.22	5.65	340.00	0.84	58.70	593.76	15.3	0.86	48.10	14.56	106.80
10	0.24	5.60	325.00	0.83	73.00	607.62	14.7	0.86	45.00	13.77	118.00
5 wt.% NaCl solution containing 0.3 mM MBT											
1	1.26	5.74	16.18	0.87	28.59	14.49	5.66	0.84	138.03	5.41	166.62
3	1.10	5.72	14.79	0.88	22.29	12.68	5.47	0.75	233.97	5.95	256.26
5	1.02	5.68	14.35	0.88	23.52	12.39	5.24	0.76	297.87	6.03	321.39
7	0.91	5.67	13.95	0.88	21.79	11.92	5.21	0.75	344.60	6.33	366.39
10	0.86	5.58	13.49	0.89	17.25	11.20	5.31	0.76	348.75	6.45	366.00
5 wt.% NaCl solution containing 0.3 mM MBT with the addition of 0.1 wt.% KI											
1	2.52	5.47	14.89	0.88	21.26	12.70	6.24	0.84	113.00	5.83	134.26
3	1.68	5.56	12.71	0.89	10.52	9.87	6.07	0.69	251.30	7.34	261.82
5	1.30	5.59	12.28	0.89	14.11	9.91	5.86	0.72	299.30	7.33	313.41
7	1.15	5.59	11.90	0.90	12.26	9.50	5.91	0.70	338.50	8.01	350.76
10	1.28	5.60	11.58	0.90	12.84	9.32	5.85	0.70	407.40	8.52	420.24
5 wt.% NaCl solution containing 0.3 mM MBT with the addition of 1.0 wt.% KI											
1	2.93	5.05	17.9	0.87	14.86	14.80	5.50	0.87	81.80	4.86	96.66
3	2.45	5.03	15.6	0.88	7.65	11.79	6.23	0.76	177.20	6.43	184.85
5	2.50	5.02	14.5	0.89	5.94	10.65	6.48	0.74	257.20	7.74	263.14
7	2.49	5.03	14.0	0.89	5.84	10.33	6.74	0.74	305.20	8.72	311.04
10	2.48	5.04	13.3	0.90	5.35	9.80	6.89	0.73	363.50	9.69	368.85

The polarization resistance (R_p) values, calculated as the sum of all resistances (Table 2), excluding R_Ω, are presented in Figure 6 as a function of the immersion time. The R_p values obtained for the AA6082 aluminium alloy samples immersed in a 5 wt.% NaCl solution, containing 0.3 mM MBT, with and without the addition of 0.1 wt.% KI, at 25 °C, are compared with the R_p values of the same samples, under the same conditions, immersed only in 5 wt.% NaCl solution, as reported previously [15]. A higher R_p value indicates the lower susceptibility of the AA6082 aluminium alloy sample to general corrosion. Figure 6 shows that at all immersion times a more resistive system is obtained with the addition of 0.3 mM MBT compared with the system containing no corrosion inhibitor. For the system containing 0.3 mM MBT, a continuous increase in R_p values is observed for up to 7 h of immersion, and then after 10 h of immersion the resistivity of the system does not significantly change (Figure 6). The same trend in the change of R_p values with increasing immersion time was found for the system containing 0.3 mM MBT with the addition of 0.1 wt.% KI for up to 7 h of immersion. However, the resistivity of this system increased from 7 to 10 h of immersion. For this system, the highest R_p value was obtained after 10 h of immersion. Table 2 and Figure 6 show that the addition of 0.1 wt.% KI to the inhibited system has no significant influence on the R_p values obtained from fitting the EIS response from the first to the seventh hour of immersion, while only a slight increase of the the R_p

value is observed after 10 h of immersion. The increase of the KI concentration from 0.1 to 1.0 wt.% did not further improve the corrosion resistance of the AA6082 aluminium samples immersed in 5 wt.% NaCl solution containing 0.3 mM MBT. Figure 6 shows that lower R_p values compared to both the inhibited systems with and without 0.1 wt.% KI addition were obtained from the first to the seventh hour of immersion.

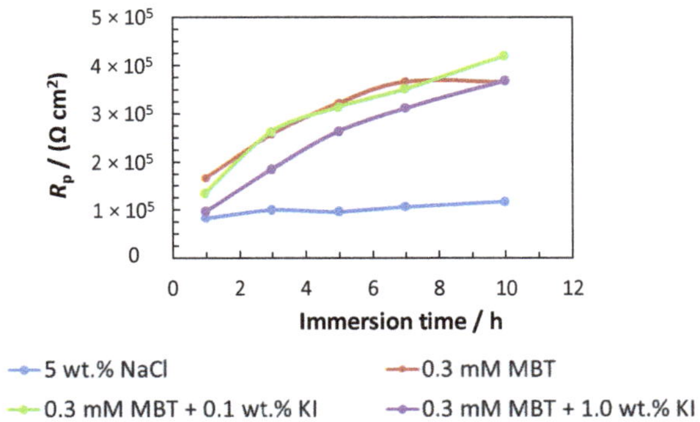

Figure 6. The variation of R_p values with increasing immersion time for the AA6082 aluminium alloy samples immersed for 1–10 h at 25 °C in 5 wt.% NaCl solution with or without 0.3 mM MBT or 0.3 mM MBT and the addition of either 0.1 or 1.0 wt.% KI.

The corrosion inhibition effectiveness of the inhibited systems, with or without the addition of 0.1 and 1.0 wt.% KI, after 10 h of immersion at 25 °C were calculated from the R_p values presented in Table 2. The corrosion inhibition effectiveness for the system containing 0.3 mM MBT was 67.76%, while the addition of 0.1 wt.% KI to the 0.3 mM MBT slightly increased the corrosion inhibition effectiveness to 71.92%. The non-significant difference in the R_p values of the inhibited samples with and without 1.0 wt.% KI addition is also reflected in the corrosion inhibition effectiveness value. The corrosion inhibition effectiveness of the samples immersed in 5 wt.% NaCl solution containing 0.3 mM MBT with 1.0 wt.% KI addition was 68.01%.

3.1.3. Potentiodynamic Curve Measurements

The potentiodynamic curves of the AA6082 aluminium alloy samples after 11 h of immersion in 5 wt.% NaCl solution with or without 0.3 mM MBT or 0.3 mM MBT and the addition of 0.1 wt.% KI are presented in Figure 7. The same as found in the open circuit potential measurements for E_{corr}, the E_{oc} of the AA6082 aluminium alloy samples shifted to more negative potentials upon the addition of 0.3 mM MBT or 0.3 mM MBT and 0.1 wt.% KI. The latter explains that 0.3 mM MBT and a combination of 0.3 mM MBT and 0.1 wt.% KI primarily inhibit the cathodic reaction of the corrosion couple, acting as a cathodic-type corrosion inhibitor (the same was concluded above by means of open circuit potential measurements). Figure 7 shows that for the inhibited samples (both with and without the addition of KI), the corrosion current density shifted to lower values compared with the non-inhibited sample. For the non-inhibited samples, the breakdown potential, E_{bd} (the potential at which a sudden increase in the current density is observed), is not clearly expressed, whereas for both additions, the E_{bd} is expressed. Moreover, based on the increased $E_{bd}-E_{oc}$ potential difference upon the addition of either 0.3 mM MBT or a combination of 0.3 mM MBT and 0.1 wt.% KI, the localized corrosion of the AA6082 aluminium alloy samples was mitigated compared with the non-inhibited sample [45].

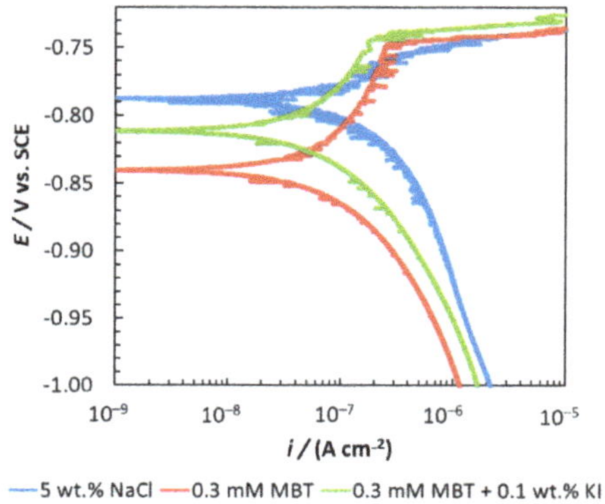

Figure 7. Potentiodynamic curves for the AA6082 aluminium alloy samples after 11 h of immersion at 25 °C in 5 wt.% NaCl solution containing 0.3 mM MBT or a combination of 0.3 mM MBT and 0.1 wt.% KI.

3.2. Weight Loss Measurements

The AA6082 aluminium alloy samples were immersed for 7 days at 25 °C in 5 wt.% NaCl solution containing 0.3 mM MBT, with and without the addition of KI as an intensifier. After the immersion time and cleaning procedure as described above, the AA6082 aluminium alloy samples showed a higher mass than before immersion, due to the inability to remove the corrosion products from the surface of the AA6082 aluminium alloy samples, even though corrosion products were visible on the surface. A cleaning procedure in acids was not performed as this could also remove the base aluminium alloy, resulting in misleading results. Similar strongly attached corrosion products were also found when the samples were immersed for 30 days and 180 days at 25 °C or 7 days at 50 °C. Therefore, for the present corrosion inhibition research weight loss measurements are not appropriate for evaluating the corrosion effectiveness of the tested corrosion inhibitors.

3.3. Surface Analysis

The AA6082 aluminium alloy samples prepared as described in Section 2.1 were subjected to surface characterization, including ATR-FTIR, XPS, ToF-SIMS, and contact angle measurements.

ATR-FTIR spectra are presented in Figure 8. In the spectra for the non-inhibited and the inhibited (with and without the addition of KI) samples, two peaks were observed at 2360 and 2336 cm^{-1}, which are connected with the asymmetric stretching of CO_2 [46]. Several other peaks were found in the inhibited samples. It was previously reported [47] that the peak at 1463 cm^{-1} is attributed to the C–N stretching vibrations. Rai et al. [48] attributed the peak at 1428 cm^{-1} to the vibrations of the C–N–H group in MBT. The peak observed at 1318 cm^{-1} has been previously attributed to the aromatic C–H in-plane bending modes [49,50]. Li et al. [51] reported that the peak at 1242 cm^{-1} is attributable to C–C stretching vibrations, while the peak at 1011 cm^{-1} is attributable to the bending C–C–C vibrations. Meanwhile, Rai et al. [48] attributed the peak at 1015 cm^{-1} not only to the C–C–C bending vibrations, but also to the C–S stretching vibrations. In addition, the ATR-FTIR spectra of the pure solid MBT chemical (powder) is also presented in Figure 8 for comparison. All the peaks presented above for the inhibited samples are also found in the spectra of the pure solid MBT. Because no characteristic peaks related to the adsorption of organic compounds were observed for the non-inhibited AA6082 aluminium alloy sample, and because of the peaks found on the inhibited samples, it can be concluded that MBT was adsorbed on the aluminium alloy samples. Moreover, the addition of 0.1 wt.% KI increased the amount of the adsorbed MBT on the surface as more signal was obtained for the latter compared to the system containing only MBT.

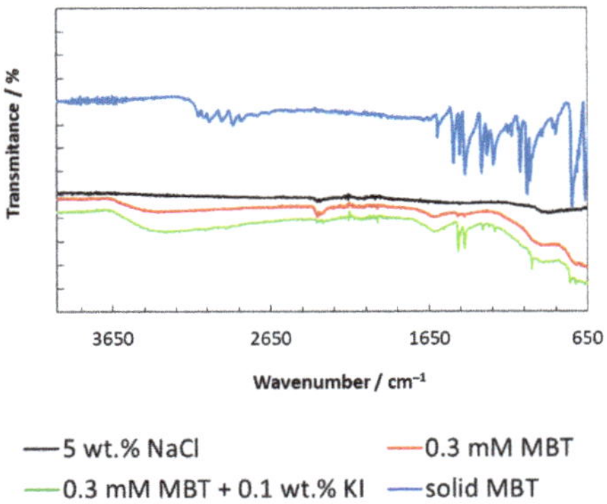

Figure 8. ATR-FTIR spectra of the AA6082 aluminium alloy samples immersed for 31 days in 5 wt.% NaCl solution with the addition of 0.3 mM MBT or a combination of 0.3 mM MBT and 0.1 wt.% KI. The ATR-FTIR spectra of the pure solid MBT is given also for comparison.

In order to confirm the adsorption of MBT with even more confidence, XPS and ToF-SIMS measurements were performed. A clear indication that MBT molecules were adsorbed on the surface is the high-resolution S 2*p* peak present for the inhibited samples (Figure 9a). Therefore, S-containing species are on the surface (in the present case, the S signal can only originate from the MBT molecule). Moreover, the negative ion ToF-SIMS spectra analysis confirmed the presence of $C_7H_4NS_2^-$ at a mass-to-charge (*m*/*z*) ratio of 165.98 (Figure 9b), representing the parent ion molecule of MBT after the removal of one proton ($(M–H)^-$ ion).

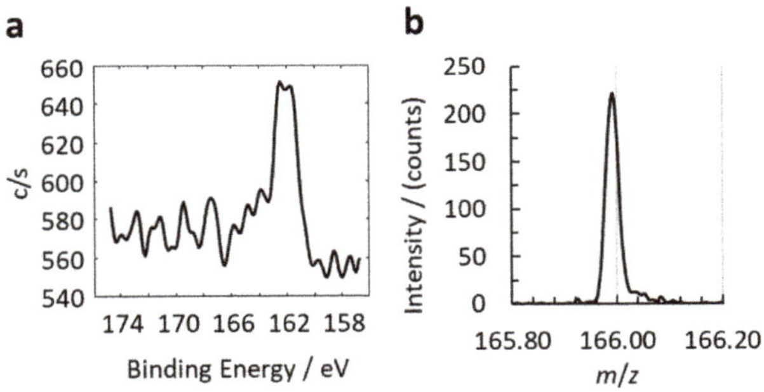

Figure 9. High-resolution (**a**) S 2*p* XPS spectra and (**b**) time-of-flight secondary ion mass spectrometry (ToF-SIMS) measurements for the MBT adsorbed on AA6082 aluminium alloy.

The influence of the adsorption of MBT on the hydrophobicity of the non-inhibited AA6082 aluminium alloy samples immersed for 1 h in 5 wt.% solution with and without the addition of 0.1 wt.% KI was studied using contact angle measurements. The average contact angles and the respective shapes of the drops for each of the tested systems at 25 °C are presented in Table 3. The highest contact angle (the most hydrophobic nature) was observed for the ground AA6082 aluminium alloy samples immersed in 5 wt.% NaCl solution (the average value for three replicate measurements was 102.5°). A significantly more hydrophilic nature was observed for the ground AA6082 aluminium alloy samples immersed in 5 wt.% NaCl solution containing 0.3 mM MBT. The average contact angle value measured for the samples immersed in the system containing 0.3 mM MBT was found to be 45.9°. Only a slight increase in the average contact angle value (51.6°) was measured for the ground AA6082 aluminium alloy samples immersed in a 5 wt.% solution containing 0.3 mM MBT with the addition of 0.1 wt.% KI. Therefore, it can be concluded that the addition of MBT renders the AA6082 aluminium alloy samples more hydrophilic. The latter suggests that the MBT surface layer does not act as a barrier for a corrosion medium, but rather forms an adsorbed layer that mitigates the oxidation (and consequently corrosion) of the AA6082 aluminium alloy samples.

Table 3. Average contact angle values (out of three replicate measurements) and the shape of the respective drop (one example) for the ground AA6082 aluminium alloy samples immersed for 1 h in 5 wt.% NaCl solution containing 0.3 mM MBT or a combination of 0.3 mM MBT and the addition of 0.1 wt.% KI at 25 °C.

Medium	Average Contact Angle [°]	Shape of the Drops on the Sample
5 wt.% NaCl solution	102.5	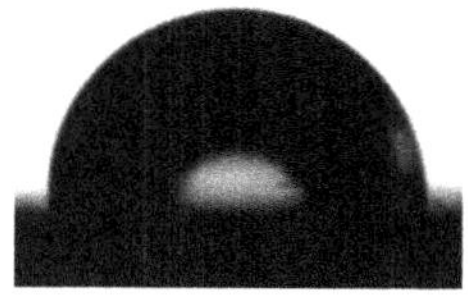
5 wt.% NaCl solution containing 3 mM MBT	45.9	
5 wt.% NaCl solution containing 3 mM MBT + 0.1 wt.% KI	51.6	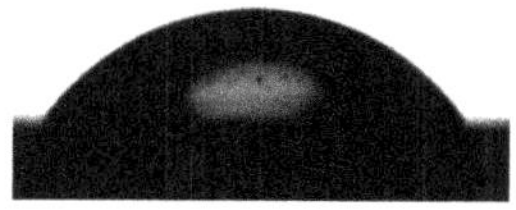

4. Conclusions

The corrosion inhibition effect of five azole compounds (i.e., 2-mercaptobenzothiazole (MBT), 2-mercaptobenzoxazole (MBO), 2-mercaptobenzimidazole (MBI), benzotriazole (BTA), and 3-amino1H-1,2,4-triazole (3-AT)) on the corrosion of AA6082 aluminium alloy samples immersed in 5 wt.% NaCl solution at 25 and 50 °C was investigated. A 5 wt.% NaCl solution at a temperature of 50 °C is usually employed in automotive corrosion tests to represent a highly corrosive environment.

The EIS measurements indicated that only MBT and 3-AT showed a corrosion inhibition effect at 50 °C. These compounds were afterwards tested at 25 °C, at which only MBT showed a corrosion inhibition effect and was, therefore, analysed further. It was shown that the corrosion of the AA6082 aluminium alloy samples in 5 wt.% NaCl solution with 0.3 mM MBT at 25 °C is only under kinetic control. EIS measurements showed that for the AA6082 aluminium alloy samples immersed in 5 wt.% NaCl solution containing 0.3 mM MBT, the surface layer (comprising the oxide layer and the adsorbed inhibitor) increased with an increase in immersion time. The susceptibility of the AA6082 aluminium alloy samples to general corrosion under these conditions decreased with increasing immersion time. Potentiodynamic curve measurements showed that the addition of MBT lowered the corrosion current density of the AA6082 aluminium alloy samples. MBT acted as a cathodic-type corrosion inhibitor by mainly inhibiting the cathodic reaction of the corrosion couple. The inhibited samples showed lower susceptibility to localized (pitting) corrosion. The addition of KI as a possible intensifier did not significantly further influence the samples' general and localized corrosion resistance, as determined by the EIS and potentiodynamic curve measurements.

The adsorption of MBT on the surface of the AA6082 aluminium alloy samples was proven by attenuated total reflectance Fourier transform infrared spectroscopy measurements, X-ray photoelectron spectroscopy, and time-of-flight secondary ion mass spectrometry. Contact angle measurements showed a significant decrease in the hydrophobicity of the samples upon the addition of 0.3 mM MBT, suggesting that the corrosion inhibition effect of MBT is not due to the formation of a barrier layer, but to a adsorbed layer that mitigates the oxidation of the base alloy.

In this study, MBT was, therefore, shown to be a corrosion inhibitor for AA6082 aluminium alloy in 5 wt.% NaCl solution. An appropriate agent that can increase the solubility of this compound will be needed to further increase its corrosion inhibition effectiveness. Furthermore, some other intensifier should be employed to increase the performance of MBT. Therefore, future studies should be oriented towards corrosion inhibitor formulation design based on MBT as a corrosion inhibitor in such mixtures.

Author Contributions: Conceptualization, M.F.; Methodology, M.F. and K.X.; Software, M.F. and K.X.; Validation, M.F. and K.X.; Formal Analysis, M.F. and K.X.; Investigation, K.X. and M.F.; Resources, M.F.; Data Curation, K.X. and M.F.; Writing—Original Draft Preparation, K.X. and M.F.; Writing—Review and Editing, K.X. and M.F.; Visualization, M.F. and K.X.; Supervision, M.F.; Project Administration, M.F.; Funding Acquisition, M.F.

Funding: This research was funded by the Slovenian Research Agency (P2-0032).

Conflicts of Interest: The authors declare no conflict of interest.

References

1. Davis, J.R. *Corrosion: Understanding the Basics*; ASM International: Novelty, OH, USA, 2000.
2. Revie, R.W. *Uhlig's Corrosion Handbook*, 3rd ed.; John Willey & Sons Inc.: New York, NY, USA, 2011.
3. Davis, J.R. *Corrosion of Aluminium and Aluminium Alloys*; ASM International: Novelty, OH, USA, 1999.
4. Poznak, A.; Freiberg, D.; Sanders, P. Chapter 10—Automotive wrought aluminium alloys. In *Fundamentals of Aluminium Metallurgy*; Lumley, R.N., Ed.; Woodhead Publishing: Sawston, UK, 2018; pp. 333–386.
5. Larsen, M.H.; Walmsley, J.C.; Lunder, O.; Mathiesen, R.H.; Nisancioglu, K. Intergranular corrosion of copper-containing AA6xxx AlMgSi aluminum alloys. *J. Electrochem. Soc.* **2008**, *155*, C550–C556. [CrossRef]
6. Zhan, H.; Mol, J.M.C.; Hannour, F.; Zhuang, L.; Terryn, H.; de Wit, J.H.W. The influence of copper content on intergranular corrosion of model AlMgSi(Cu) alloys. *Mater. Corros.* **2008**, *59*, 670–675. [CrossRef]
7. Svenningsen, G.; Lein, J.E.; Bjørgum, A.; Nordlien, J.H.; Yu, Y.; Nisancioglu, K. Effect of low copper content and heat treatment on intergranular corrosion of model AlMgSi alloys. *Corros. Sci.* **2006**, *48*, 226–242. [CrossRef]
8. Zou, Y.; Liu, Q.; Jia, Z.; Xing, Y.; Ding, L.; Wang, X. The intergranular corrosion behavior of 6000-series alloys with different Mg/Si and Cu content. *Appl. Surf. Sci.* **2017**, *405*, 489–496. [CrossRef]
9. Kairy, S.K.; Rometsch, P.A.; Diao, K.; Nie, J.F.; Davies, C.H.J.; Birbilis, N. Exploring the electrochemistry of 6xxx series aluminium alloys as a function of Si to Mg ratio, Cu content, ageing conditions and microstructure. *Electrochim. Acta* **2016**, *190*, 92–103. [CrossRef]
10. Liang, W.J.; Rometsch, P.A.; Cao, L.F.; Birbilis, N. General aspects related to the corrosion of 6xxx series aluminium alloys: Exploring the influence of Mg/Si ratio and Cu. *Corros. Sci.* **2013**, *76*, 119–128. [CrossRef]
11. Deepa, P.; Padmalatha, R. Corrosion behaviour of 6063 aluminium alloy in acidic and in alkaline media. *Arab. J. Chem.* **2017**, *10*, S2234–S2244. [CrossRef]
12. Cicolin, D.; Trueba, M.; Trasatti, S.P. Effect of chloride concentration, pH and dissolved oxygen, on the repassivation of 6082-T6 Al alloy. *Electrochim. Acta* **2014**, *124*, 27–35. [CrossRef]
13. Panagopoulos, C.N.; Georgiou, E.; Giannakopoulos, K.I.; Orfanos, P.G. Effect of pH on stress corrosion cracking of 6082 Al alloy. *Metals* **2018**, *8*, 578. [CrossRef]
14. Trdan, U.; Grum, J. Evaluation of corrosion resistance of AA6082-T651 aluminium alloy after laser shock peening by means of cyclic polarisation and EIS methods. *Corros. Sci.* **2012**, *59*, 324–333. [CrossRef]
15. Xhanari, K.; Finšgar, M. Electrochemical analysis of AA6082 aluminium alloy in chloride media. *Int. J. Electrochem. Sci.* **2017**, *12*, 5845–5853. [CrossRef]
16. Firouzi, A.; Del Gaudio, C.; lamastra, F.R.; Montesperelli, G.; Bianco, A. Electrospun polymeric coatings on aluminum alloy as a straightforward approach for corrosion protection. *J. Appl. Polym. Sci.* **2015**, *132*, 41250. [CrossRef]
17. Mori, S.; Lamastra, F.R.; Kaciulis, S.; Soltani, P.; Montesperelli, G. Low-temperature titania coatings for aluminium corrosion protection. *Corros. Eng. Sci. Technol.* **2018**, *53*, 44–53. [CrossRef]
18. Panagopoulos, C.N.; Georgiou, E.P.; Gavras, A.G. Corrosion and wear of 6082 aluminum alloy. *Tribol. Int.* **2009**, *42*, 886–889. [CrossRef]
19. Xhanari, K.; Finšgar, M. Organic corrosion inhibitors for aluminium and its alloys in acid solutions: A review. *RSC Adv.* **2016**, *6*, 62833–62857. [CrossRef]

20. Xhanari, K.; Finšgar, M. Organic corrosion inhibitors for aluminum and its alloys in chloride and alkaline solutions: A review. *Arab. J. Chem.* **2016**, in press. [CrossRef]
21. Xhanari, K.; Finsgar, M.; Knez Hrncic, M.; Maver, U.; Knez, Z.; Seiti, B. Green corrosion inhibitors for aluminium and its alloys: A review. *RSC Adv.* **2017**, *7*, 27299–27330. [CrossRef]
22. Xhanari, K.; Finsgar, M. The first electrochemical and surface analysis of 2-aminobenzimidazole as a corrosion inhibitor for copper in chloride solution. *New J. Chem.* **2017**, *41*, 7151–7161. [CrossRef]
23. Finšgar, M. 2-mercaptobenzimidazole as a copper corrosion inhibitor: Part I. Long-term immersion, 3D-profilometry, and electrochemistry. *Corros. Sci.* **2013**, *72*, 82–89. [CrossRef]
24. Finšgar, M. Electrochemical analysis of 4-methyl-2-phenyl-imidazole adsorbed on Cu. *Int. J. Electrochem. Sci.* **2016**, *11*, 6775–6790. [CrossRef]
25. Finšgar, M.; Merl, D.K. 2-Mercaptobenzoxazole as a copper corrosion inhibitor in chloride solution: Electrochemistry, 3D-profilometry, and XPS surface analysis. *Corros. Sci.* **2014**, *80*, 82–95. [CrossRef]
26. Finšgar, M.; Merl, D.K. An electrochemical, long-term immersion, and XPS study of 2-mercaptobenzothiazole as a copper corrosion inhibitor in chloride solution. *Corros. Sci.* **2014**, *83*, 164–175. [CrossRef]
27. Finšgar, M.; Lesar, A.; Kokalj, A.; Milošev, I. A comparative electrochemical and quantum chemical calculation study of BTAH and BTAOH as copper corrosion inhibitors in near neutral chloride solution. *Electrochim. Acta* **2008**, *53*, 8287–8297. [CrossRef]
28. Finšgar, M.; Xhanari, K. Electrochemical and surface analysis of 2-phenylimidazole adsorbed on copper from chloride solution. *Coatings* **2018**, *8*, 234. [CrossRef]
29. Huang, H.; Wang, Z.; Gong, Y.; Gao, F.; Luo, Z.; Zhang, S.; Li, H. Water soluble corrosion inhibitors for copper in 3.5 wt % sodium chloride solution. *Corros. Sci.* **2017**, *123*, 339–350. [CrossRef]
30. Curkovic, H.O.; Stupnisek-Lisac, E.; Takenouti, H. The influence of pH value on the efficiency of imidazole based corrosion inhibitors of copper. *Corros. Sci.* **2010**, *52*, 398–405. [CrossRef]
31. Petrović Mihajlović, M.B.; Radovanović, M.B.; Tasić, Ž.Z.; Antonijević, M.M. Imidazole based compounds as copper corrosion inhibitors in seawater. *J. Mol. Liquids* **2017**, *225*, 127–136. [CrossRef]
32. Mennucci, M.M.; Banczek, E.P.; Rodrigues, P.R.P.; Costa, I. Evaluation of benzotriazole as corrosion inhibitor for carbon steel in simulated pore solution. *Cem. Concr. Compos.* **2009**, *31*, 418–424. [CrossRef]
33. Finšgar, M.; Petovar, B.; Xhanari, K.; Maver, U. The corrosion inhibition of certain azoles on steel in chloride media: Electrochemistry and surface analysis. *Corros. Sci.* **2016**, *111*, 370–381. [CrossRef]
34. Sabet Bokati, K.; Dehghanian, C.; Yari, S. Corrosion inhibition of copper, mild steel and galvanically coupled copper-mild steel in artificial sea water in presence of 1H-benzotriazole, sodium molybdate and sodium phosphate. *Corros. Sci.* **2017**, *126*, 272–285. [CrossRef]
35. Abd El Haleem, S.M.; Abd El Wanees, S.; Bahgat, A. Environmental factors affecting the corrosion behaviour of reinforcing steel. VI. Benzotriazole and its derivatives as corrosion inhibitors of steel. *Corros. Sci.* **2014**, *87*, 321–333. [CrossRef]
36. Sherif, E.-S.M. Electrochemical investigations on the corrosion inhibition of aluminum by 3-amino-1,2,4-triazole-5-thiol in naturally aerated stagnant seawater. *J. Ind. Eng. Chem.* **2013**, *19*, 1884–1889. [CrossRef]
37. Zor, S.; Sağdinç, S. Experimental and theoretical study of sulfathiazole as environmentally friendly inhibitor on aluminum corrosion in NaCl. *Prot. Met. Phys. Chem. Surf.* **2014**, *50*, 244–253. [CrossRef]
38. Harvey, T.G.; Hardin, S.G.; Hughes, A.E.; Muster, T.H.; White, P.A.; Markley, T.A.; Corrigan, P.A.; Mardel, J.; Garcia, S.J.; Mol, J.M.C.; et al. The effect of inhibitor structure on the corrosion of AA2024 and AA7075. *Corros. Sci.* **2011**, *53*, 2184–2190. [CrossRef]
39. Lamaka, S.V.; Zheludkevich, M.L.; Yasakau, K.A.; Montemor, M.F.; Ferreira, M.G.S. High effective organic corrosion inhibitors for 2024 aluminium alloy. *Electrochim. Acta* **2007**, *52*, 7231–7247. [CrossRef]
40. Coelho, L.B.; Cossement, D.; Olivier, M.-G. Benzotriazole and cerium chloride as corrosion inhibitors for AA2024-T3: An EIS investigation supported by SVET and ToF-SIMS analysis. *Corros. Sci.* **2018**, *130*, 177–189. [CrossRef]
41. *ISO 9227:2006(E) Corrosion Tests in Artificial Atmospheres—Salt Spray Tests*; ISO: Geneva, Switzerland, 2006.
42. Finšgar, M.; Jackson, J. Application of corrosion inhibitors for steels in acidic media for the oil and gas industry: A review. *Corros. Sci.* **2014**, *86*, 17–41. [CrossRef]
43. Massart, D.L.; Vandeginste, B.G.M.; Buydens, L.M.C.; Jong, S.D.; Lewi, P.J.; Smeyers-Verbeke, J. *Handbook of Chemometrics and Qualimetrics: Part A*; Elsevier Science: Amsterdam, The Netherlands, 1997; p. 112.

44. Kovač, J.; Finšgar, M. Analysis of the thermal stability of very thin surface layers of corrosion inhibitors by time-of-flight secondary ion mass spectrometry. *J. Am. Soc. Mass Spectrom.* **2018**, *29*, 2305–2316. [CrossRef]
45. Finšgar, M.; Milošev, I. Corrosion behaviour of stainless steels in aqueous solutions of methanesulfonic acid. *Corros. Sci.* **2010**, *52*, 2430–2438. [CrossRef]
46. Gerakines, P.A.; Schutte, W.A.; Greenberg, J.M.; van Dishoeck, E.F. The infrared band atrengths of H_2O, CO and CO_2 in laboratory simulations of astrophysical ice mixtures. *Astron. Astrophys.* **1995**, *296*, 810–818.
47. Güllüoğlu, M.T.; Erdogdu, Y.; Karpagam, J.; Sundaraganesan, N.; Yurdakul, Ş. DFT, FT-Raman, FT-IR and FT-NMR studies of 4-phenylimidazole. *J. Mol. Struct.* **2011**, *990*, 14–20. [CrossRef]
48. Rai, A.K.; Singh, R.; Singh, K.N.; Singh, V.B. FTIR, Raman spectra and ab initio calculations of 2-mercaptobenzothiazole. *Spectrochim. Acta Part A Mol. Biomol. Spectrosc.* **2006**, *63*, 483–490. [CrossRef] [PubMed]
49. Sudha, S.; Karabacak, M.; Kurt, M.; Cinar, M.; Sundaraganesan, N. Molecular structure, vibrational spectroscopic, first-order hyperpolarizability and HOMO, LUMO studies of 2-aminobenzimidazole. *Spectrochim. Acta Part A Mol. Biomol. Spectrosc.* **2011**, *84*, 184–195. [CrossRef] [PubMed]
50. Erdogdu, Y.; Eskioğlu, B.; Güllüoğlu, M.T. Theoretical investigations on the molecular structure and vibrational spectral analysis of 4-methyl 2-phenylimidazole. *Opt. Spectrosc.* **2012**, *113*, 596–606. [CrossRef]
51. Li, X.; Tang, Z.; Zhang, X. Molecular structure, IR spectra of 2-mercaptobenzothiazole and 2-mercaptobenzoxazole by density functional theory and ab initio Hartree–Fock calculations. *Spectrochim. Acta Part A Mol. Biomol. Spectrosc.* **2009**, *74*, 168–173. [CrossRef] [PubMed]

coatings

MDPI

Article

Electrochemical Impedance Spectroscopy Investigation on the Corrosive Behaviour of Waterborne Silicate Micaceous Iron Oxide Coatings in Seawater

Xu Zhao [1], Yuhong Qi [1,*], Zhanping Zhang [1], Kejiao Li [1] and Zhitao Li [2]

1 Department of Materials Science and Engineering, Dalian Maritime University, Dalian 116026, China
2 Dalian Jinzhong (Lvshun) Heavy Industry Co., Ltd., Dalian 116041, China
* Correspondence: yuhong_qi@dlmu.edu.cn

Received: 17 May 2019; Accepted: 26 June 2019; Published: 28 June 2019

Abstract: The anticorrosive composite coatings based on waterborne silicate were prepared to replace existing solvent-based coatings suitable for ships. A series of composite coatings were prepared by adding zinc powder and micaceous iron oxide to the waterborne silicate resin. The adhesion, pencil hardness and impact resistance of the coatings were investigated and corrosion performance in seawater is characterized by electrochemical impedance spectroscopy (EIS). The results show that coatings have excellent adhesion and impact resistance and their pencil hardness can reach up to 4H. During the immersion of composite coatings in seawater for 8 h, only one time constant appears in the Bode plot, coating capacitance (Q_c) gradually increases but dispersion coefficient (n) and coating resistance (R_c) gradually decrease. The breakpoint frequency formula was deduced, considering the dispersion effect. With the increase of micaceous iron oxide, the fluctuation of breakpoint frequency with immersion time is weakened. It can be used to evaluate the corrosion resistance of inorganic anticorrosive coatings in seawater. In addition, different penetration models of corrosive media were proposed for the coatings with low or high content of micaceous iron oxide.

Keywords: waterborne inorganic coating; anticorrosion coating; micaceous iron oxide; EIS; breakpoint frequency

1. Introduction

A ship's corrosion in the ocean is unavoidable. Therefore, its corrosion rate must be limited to extend the life of ship in service [1]. There are two main methods for ship corrosion protection: coating and cathodic protection. Cathodic protection includes sacrificial anode and impressed current protection. The former protects the hull with a metal or alloy that is more negative than the electrode potential of steel. The timeliness of this method is not well controlled, and the application position is limited. The latter is to apply a protection current to the hull through the auxiliary anode supported by a DC power source. This method has a high investment at one time, complex installation and could be applied to hulls. Compared to cathodic protection, coating protection is an economical, convenient and less restrictive method. The coating protection relies mainly on the anti-corrosion coating used. The more mature and extensive coatings currently in use are solvent-based anti-corrosion coatings but they will release volatile organic compounds (VOC), which results in photochemical smog, acid rain and serious pollution to the atmosphere. VOC are considered one of the major sources of atmospheric pollution. As the awareness of environmental and labour protection of society is growing, some countries around the world have successively established laws and regulations that limit VOC emissions. Therefore, the research and development of environmentally friendly coatings that replace traditional solvent-based coatings has become the trend. Because waterborne coatings use water as a solvent and dispersant, resulting in very low or zero VOC emissions, they are more beneficial to the human living environment and health. Waterborne coatings include both organic and inorganic. The former mainly includes waterborne epoxy, polyurethane and acrylic anticorrosion coating and so forth. Their anticorrosive properties can be improved by modified resin, pigment, filler and additive. For example, for the epoxy coating, its anticorrosion property can be improved by means of polyaniline modifying epoxy resin [2–5] or by adding nano-silica/aluminium powder together with suitable additive [6,7]. The polyurethane coating can be improved by means of epoxy/acrylic modifying polyurethane [8–11] or by adding nano-zinc oxide [12,13]. The acrylic coating can be improved by means of epoxy modifying acrylic acid [14] or by adding suitable functional monomer, pigment and filler and additive [15–17]. Compared with the solvent-based anticorrosion coating, the waterborne organic anticorrosion coating's environmental protection has been greatly improved; however, since its binder is usually composed of resins dissolved in organic solvents, it will inevitably release VOC during the manufacturing process, which has limited its development to some extent. On the contrary, the inorganic binders without the VOC are completely environmentally friendly and have become a hot spot for coating research. They mainly include phosphate [18,19], silica sol [20,21] and silicate coatings [22,23]. Silica sol paints form a film by polymerization of silanol group but cohesion is larger during polycondensation, which is likely to cause the coatings to crack or even cause large-areas to fall off. Phosphate coatings can be formed by polycondensation of phosphate at high temperatures and the ratio of metal atom to phosphorus (M/P) determines the final properties of the coatings. Silicate coatings are also referred to as water glass coatings. The general formula of water glass is $M_2O{\cdot}m SiO_2{\cdot}n H_2O$, where M is Li or Na or K and m is modulus. As the modulus increases, the bonding ability, bonding strength and water resistance of the silicate increase accordingly but solubility in water decreases. Although these inorganic coatings have excellent environmental performance but still have some problems, such as poor water resistance, film formation and mechanical properties and so forth [24]. In order to improve the performance of waterborne inorganic anticorrosion coatings, acid modification, organic-inorganic hybridization and nano-modification have been used [6]. In this paper, silicate resin modified by nano-sized silicone-acrylic emulsion was selected. Silicone-acrylic emulsion has the advantages of both silicone and acrylic resin. Silicone is used to improve the water resistance, acrylic resin to improve the mechanical property and nano-effects are used to enhance the overall performance. Micaceous iron oxide is a natural mineral and its main chemical composition is α-Fe_2O_3, which is an unusual tabular crystalline form capable of fracturing into very thin plate-like cleavage fragments called lamellae so as to produce the best protective action [25]. This is due to the fact that the lamellae can be distributed in parallel within the coating to introduce an "labyrinth effect," which makes the diffusion path of corrosive media such as water, oxygen and so forth, in the coating become more tortuous, thereby increasing the time when corrosive media reach a metal substrate, reducing the formation rate of corrosion galvanic cells and slowing the corrosion rate [24,26–28]. This has been proven in some literature. For example, Yang et al. [29] found that polymer coatings containing impermeable inorganic flakes can reduce solute permeability by a factor of 10 and Danaee et al. [30] found that the rate of reactivity of inorganic zinc-rich coatings was reduced after replacing the zinc powder with micaceous iron oxide. Micaceous iron oxide's addition not only improves the protective performance of the coatings but also reduces the curing shrinkage of the coating, thereby improving the comprehensive performance of the coatings.

Many investigations [31–33] have proven that the corrosion resistance of organic coatings could be evaluated by electrochemical parameters such as coating capacitance (Q_c), coating resistance (R_c), double-layer capacitance and charge transfer resistance obtained by fitting the electrochemical impedance spectroscopy of the coatings. This is mainly due to the fact that some assumptions, such as intactness of coating main structure, linear transmission of corrosive media and so forth, are needed to apply the corresponding relationship [34] between Q_c and the water absorption of the coating. However, these conditions cannot be met in most cases for inorganic coatings—it will result in an obvious deviation between the water absorption calculated from the coating capacitance value and the actual water absorption [35–38]. Therefore, to effectively evaluate the protective performance of the inorganic coatings, new evaluation parameters and methods need to be found. The effectiveness of the breakpoint frequency (f_*) proposed by Liu [39] has been verified for the evaluation of the protectiveness of organic coatings representing dispersion effect. In view of that, the coatings studied also show the dispersion effect in this paper. Consequently, R_c and f_* will be used for this investigation.

2. Experiment

2.1. Materials

Nanomodified waterborne silicate resin for zinc-rich paint (E777-2) was obtained from the Wuhan Modern Technologies Institute (Wuhan, China). The E777-2's main constituent is potassium silicate, its solid content, pH and density are 25%, 10–12 and 1.02–1.20 g/mL (25 °C), respectively. Spherical zinc powder through 400 mesh and 500 mesh screen and flake zinc powder were purchased from Jiangsu Kecheng Nonferrous Metal New Material Co., Ltd. (Jiangsu, China) Talc, composite calcium zinc phosphate, titanium dioxide and micaceous iron oxide was respectively purchased from Dalian Keli Ultrafine Powder Co., Ltd. (Dalian, China), Wuhan Modern Technologies Institute, Shanghai Hongyunyuan Chemical Co., Ltd. (Shanghai, China) and Changzhou Lehuan Trading Co., Ltd. (Changzhou, China). Triethylamine was bought from Tianjin Kemiou Chemical Reagent Co., Ltd. (Tianjin, China). Sterilized seawater was obtained by the process of pouring the appropriate amount of fresh seawater taken from the Yellow Sea into a funnel with filter paper, filtering seawater into a conical flask, heating the conical flask on an electric furnace and boiling the filtered seawater for more than 15 min.

2.2. Preparation of Zinc-Rich Antirust Paint

2.2.1. Formulation

The basic formulation designed for antirust paint in this experiment is shown in Table 1. ZF, Z4S and Z5S represent formulation containing flake zinc powder, 400 and 500 mesh spherical zinc powder, respectively.

Table 1. The basic formulation of zinc-rich antirust paint.

Raw Materials (wt.%)	Z4S	Z5S	ZF
E777-2	30	30	30
Spherical zinc powder (400 mesh)	70	0	0
Spherical zinc powder (500 mesh)	0	70	0
Flake zinc powder	0	0	70

2.2.2. Preparation

E777-2 was poured into a 100 mL stirring tank, the stirring speed was 500 r/min, zinc powder was added to the tank according to the amount in the formulation and the speed was increased to 1500 r/min, the mixture needed to be stirred for 15 min to prepare the required zinc-rich antirust paint. During the preparation of zinc-rich antirust paints, the formulation ZF in Table 1 was found to be incapable of forming well-dispersed paint. This is mainly due to the fact that the specific gravity of flake zinc powder is about 1/3 that of spherical zinc powder, resulting in the volume of flake zinc powder being about three times the volume of spherical zinc powder in the case of the same amount of zinc powder in the paint. The resin cannot enwrap such a large volume of flake zinc powder and therefore it is unable to make proper paint.

2.3. Preparation of Micaceous Iron Oxide Anticorrosion Paint

2.3.1. Formulation

The basic formulation designed for anticorrosion paint in this experiment is shown in Table 2. M16, M14, M12, M10 and M8 represent a sample with 16%, 14%, 12%, 10% and 8% micaceous iron oxide respectively.

Table 2. The basic formulation of micaceous iron oxide anticorrosion paint.

Raw Materials (wt.%)	M16	M14	M12	M10	M8
E777-2	64	64	64	64	64
Talc	4	4	4	4	4
Composite calcium zinc phosphate	5	5	5	5	5
Titanium dioxide	1	1	1	1	1
Deionized water	10	12	14	16	18
Micaceous iron	16	14	12	10	8

2.3.2. Preparation

Firstly, E777-2 and deionized water was added into a 300 mL stirring tank, then four kinds of powders were added in turn and stirred at 1500 r/min for 2 min after each powder was added, finally an appropriate amount of triethylamine was added and the mixture was stirred at 1500 r/min for 15 min.

2.4. Preparation of Studied Coatings

2.4.1. Preparation of Coating Specimen

In this experiment, the film thickness was controlled by brushing different layers of paint on the steel plate (Table 3). The film thickness could reach about 20, 30 and 40 μm, respectively, when the paint was brushed to two, four and six layers. Z4S-2, Z4S-3 and Z4S-4 represents formulation Z4S sample with 20, 30 and 40 μm film thickness, respectively. Z5S-2, Z5S-3 and Z5S-4 represent the formulation of the Z5S sample with film thicknesses of 20, 30 and 40 μm film, respectively.

Table 3. Film thicknesses of zinc-rich antirust paints with different formulations.

Sample	Coating Number	Dry Film Thickness (μm)
	2	19.3
Z4S	4	28.6
	6	40.2
	2	18.7
Z5S	4	30.8
	6	38.7

Each paint was brushed on the surface of aluminium plates for basic mechanical properties testing. The zinc-rich antirust paints prepared in Section 2.2.2 were brushed on the steel plate and then micaceous iron oxide anticorrosion paints were brushed on the zinc-rich antirust coating to prepare the composite coatings for electrochemical testing in seawater. Each paint was cured for 3 days.

2.4.2. Curing Mechanism of Silicate Paints

The film curing mechanism of silicate coatings has been introduced in the relevant literature [40–42]. It is generally agreed that its curing process consists of the following reaction equations. The substrate can react with silicic acid to form iron silicate to make the coating have good adhesion and the reaction of zinc with silicic acid and the polycondensation between silicic acid ensure the film formation of the coating.

$$K_2O \cdot mSiO_2 \cdot nH_2O + (2m - n + 1)H_2O \rightleftharpoons 2KOH + mSi(OH)_4, \quad (1)$$

$$2KOH + CO_2 + H_2O \rightarrow K_2CO_3 + 2H_2O, \quad (2)$$

$$M + H_2SiO_3 \rightarrow MSiO_3 + H_2, \quad (3)$$

$$mSi(OH)_4 \rightarrow HO(SiO_2)_mH + (2m - 1)H_2O. \quad (4)$$

where M represents Fe or Zn.

2.5. Preparation of Cross Section Specimens

A sample with a size of 20 × 20 mm^2 was removed from the coating Z5S + M10/Z5S + M14 with a saw blade and then the sample was fixed with a steel splint. The non-cutting edge of the sample was sanded by 400 to 1000 grit sandpaper and polished.

2.6. Measurement and Characterization

2.6.1. Confocal Laser Scanning Microscope (CLSM)

The particle size of zinc powder, surface roughness of the zinc-rich coating and cross section of the coating Z5S + M10/Z5S + M14 were measured using the LEXT software of Olympus OLS4000 CLSM version 2.2.4.

2.6.2. Dry Film Thickness

Five points were measured for each sample with a film thickness gauge (QuaNix7500, AUTOMATION DR.NIX GmbH KÖLN, Köln, Germany).

2.6.3. Water Contact Angle

Water contact angle measurements were conducted using the sessile drop method on a JC2000 contact angle measurement system. A 3 μL droplet of distilled water was placed on the surface of the coating sample using a syringe. Digital images of the droplet silhouette were captured with a charge-couple device (CCD) camera and the contact angles were evaluated using the measuring angle method. For each sample, the mean of the contact angle measurements for the five points was calculated.

2.6.4. Basic Mechanical Properties

According to GB/T 1720-1979 [43], the adhesion of the coatings was evaluated with the spiral scoring method using the film adhesion tester (QFZ-II). According to GB/T6739-2006 [44] and GB/T1732-1993 [45], the pencil hardness and impact strength of the coatings were evaluated with a pencil-scratch hardness tester (QHQ-A) and paint film impactor (QCJ), respectively.

2.6.5. Electrochemical Impedance Spectroscopy (EIS)

At different immersion times, the ZAHNER IM6ex Electrochemical Workstation (Kronach, Germany) with COLT system was used to measure the electrochemical impedance spectroscopy (EIS) of the composite coatings immersed in sterilized seawater. The signal amplitude was 10 mV relative to the open circuit potential and the frequency ranged from 1 Hz to 100 kHz. The detailed test operation can be found in Reference [46]. The test results were treated using Z2.03 USB software and were then analysed using ZSimpWin3.2.1 software.

3. Results and Discussion

3.1. Influence of Thickness on Film Forming Effect of Zinc-Rich Coatings

It was found that the surface roughness of the coating increased as the film thickness increased (refer to Table 4). This may be due to the increase in dry film thickness, which may cause the surface and interior of the paint to not be cured at the same time, producing residual stress in the coating. In addition, according to the curing mechanism of the zinc-rich coating (Equations (1)–(4)), H_2 and H_2O will be released during the curing process. If the released gas diffuses to the surface, pores will form. If the diffusion fails, bubbles can form in the interior. Therefore, the increase in thickness of the zinc-rich coating makes it subject to the collective effect of residual stress and the released gas, which leads to a poor film forming effect and increased surface roughness.

Table 4. The surface roughness of samples with different formulations and film thicknesses.

Sample	Surface Roughness (μm)
Z4S-2	1.8
Z4S-3	2.0
Z4S-4	2.7
Z5S-2	1.7
Z5S-3	1.8
Z5S-4	1.9

3.2. Effect of Zinc Particle Size and Film Thickness on Hydrophilicity of Zinc-Rich Coatings

The morphology of the zinc powder observed by using CLSM is shown in Figure 1 and the field of view of CLSM is 131 μm × 131 μm. A number of zinc powder particles were selected on the CLSM images and their particle sizes were measured by using the LEXT software (version 2.2.4) of the OLS4000. The average particle size of 400 and 500 mesh zinc powder is 3.78 and 1.87 μm, respectively.

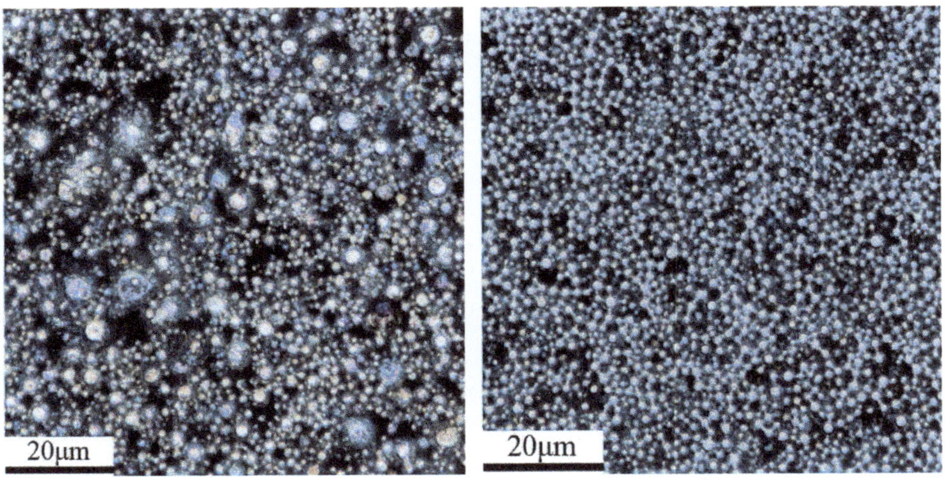

(a) (b)

Figure 1. The CLSM picture of (**a**) 400 and (**b**) 500 mesh zinc powder.

In this experiment, the water contact angles of samples with different formulations and film thicknesses were measured (Table 5) to understand the influence of zinc powder size and film thickness on the hydrophilicity. First, the water contact angles of Z4S and Z5S coatings were less than 90°, which shows that Z4S and Z5S coatings can be wetted by water. Second, when the film thickness was the same, the water contact angles of Z4S and Z5S coatings had a small difference, which shows that the zinc powder size has little effect on the hydrophilicity. Third, when the zinc powder size was constant, the apparent contact angle (θ_γ) gradually decreased as the film thickness increased. This is mainly due to the fact that the surface roughness of coatings increases as the film thickness increases, in the case where the θ_γ is less than 90°, eigen contact angle (θ) < 90° is known from the Wenzel equation [47] ($\cos\theta_\gamma = \gamma\cos\theta$, where the γ is the ratio of the actual to the geometric solid-liquid interface contact area and is ≥ 1). Wenzel [48] pointed out that when the θ is less than 90°, the θ_γ decreases as the surface roughness increases.

Table 5. Water contact angles of samples with different formulations and film thicknesses.

Sample	Water Contact Angle (°)
Z4S-2	87.8
Z4S-3	85.2
Z4S-4	65.1
Z5S-2	87.9
Z5S-3	86.2
Z5S-4	66.6

Considering the influence of film forming effect and hydrophilicity, the Z5S formulation was selected for further investigation of the composite coatings.

3.3. *Basic Mechanical Properties of Micaceous Iron Oxide Anticorrosion Coatings*

The basic mechanical properties of the micaceous iron oxide anticorrosion coatings are shown in Table 6. It can be found that the adhesion and impact strength of the coatings do not change and reach Grade 1 and 50 kg·cm, respectively. This shows that the added pigments and fillers and their contents do not affect the adhesion of the resin to the metal substrate and the enwrapping of the resin to the pigments and fillers. The paint can form a continuous film after being cured. Since the resin is bonded with the metal substrate by a chemical bond (Equation (3)), the coatings have good cohesiveness and flexibility.

Table 6. Basic mechanical properties of micaceous iron oxide anticorrosion coatings.

Sample	Pencil Hardness (H)	Adhesion (Grade)	Impact Resistance (kg·cm)
M16	4	1	50
M14	3	1	50
M12	3	1	50
M10	3	1	50
M8	2	1	50

From Table 6, it can be seen that the pencil hardness increases from 2H to 4H as the contents of micaceous iron oxide increase from 8% to 16%. This shows that the addition of micaceous iron oxide can indeed improve the scratch resistance of the coating but there is limit for the amount of micaceous iron oxide.

3.4. *Anticorrosion Performance of Composite Coatings*

3.4.1. Corrosion of Composite Coatings in Seawater

The AC impedance of the composite coatings immersed in seawater was measured at 1 h, 2 h, 4 h, 6 h and 8 h. The results for the Z5S + M16 coatings are shown in Figure 2 while the results of the other coatings are shown in Figures A1–A4

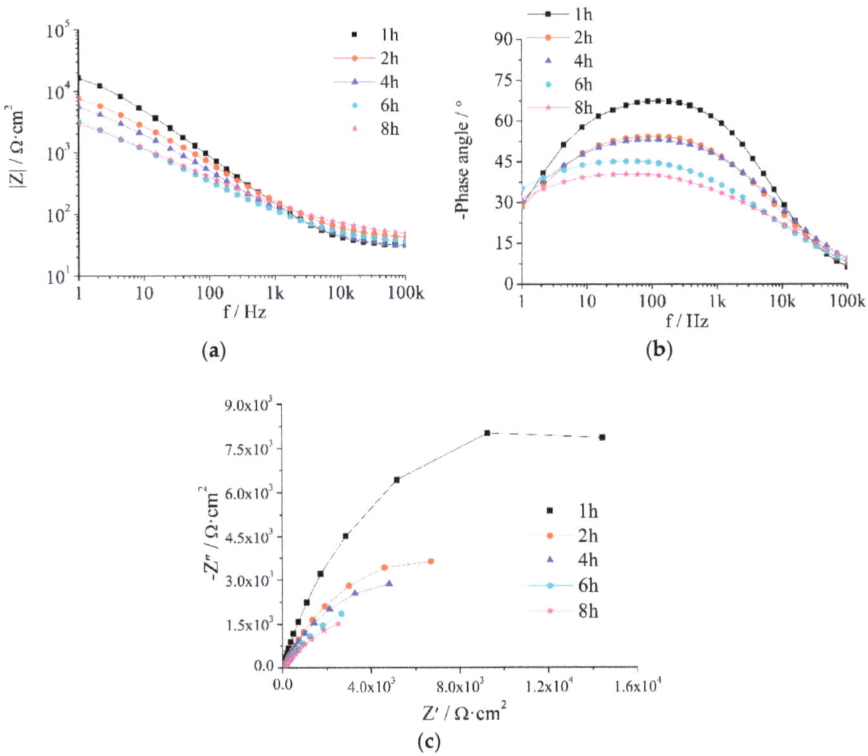

Figure 2. EIS of Z5S + M16 immersed in seawater for different times.

Based on the Bode plot in Figure 2a,b, each |Z|-*f* curve is basically a diagonal and there is only one peak on the negative phase angle-*f* curve. It shows that the composite coatings always present the impedance spectrum of one time constant during the test period; in other words, the seawater does not penetrate the anticorrosion coating and reach the antirust coating over 8 h. Compared with some conventional coatings, such as waterborne modified epoxy [49] and polyurethane coating [50], the protection performance of the composite coatings prepared in this paper was better than that of these conventional coatings.

The Nyquist plot in Figure 2c indicates that the capacitive loop decreases with an extension of the immersion time. From the Bode plot in Figure 2, it can be seen that with the prolongation of immersion time, the low frequency impedance reduces and the |Z|-*f* curve moves toward the low frequency direction, the phase angle curve descends and moves toward the high frequency direction. These show that Q_c increases while R_c decreases as immersion time increases, it indicates that corrosive media penetrate gradually into the composite coating.

The measurement results in Figure 2 were analysed using the equivalent circuit R(QR) (Figure 3) in the ZSimpWin3.2.1 software and the fitted results are shown in Table 7. Q_c gradually increases while dispersion coefficient (n) and R_c gradually decrease as immersion time is extended.

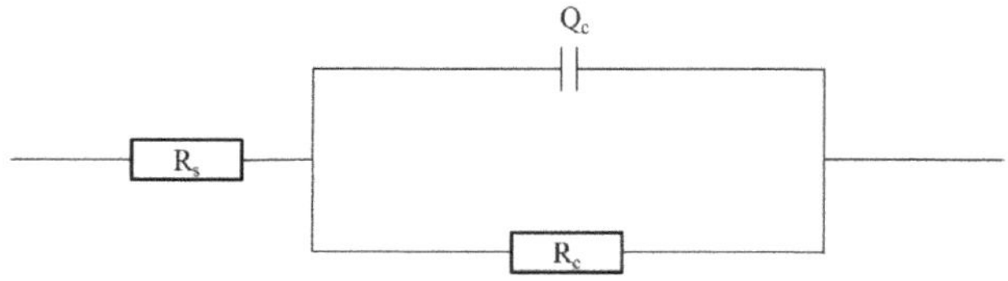

Figure 3. Equivalent circuit used to fit the EIS data.

Table 7. Electrochemical parameters of the equivalent circuit for micaceous iron oxide composite coatings immersed at different times in seawater.

Sample	Time (h)	Q_c ($\mu F \cdot cm^{-2} \cdot Hz^{1-n}$)	n	R_c ($k\Omega \cdot cm^2$)
Z5S + M8	1	6.499	0.7806	16.02
	2	10.83	0.6998	6.308
	4	14.35	0.6534	5.526
	6	23.85	0.638	5.466
	8	24.55	0.5874	5.169
Z5S+M10	1	8.873	0.7739	18.98
	2	18.04	0.716	7.462
	4	36.07	0.6433	7.209
	6	70.57	0.5695	7.193
	8	87.85	0.5147	6.76
Z5S+M12	1	8.665	0.7959	20
	2	18.31	0.6991	9.314
	4	30.95	0.6603	8.848
	6	31.63	0.6042	8.39
	8	37.4	0.5914	7.043
Z5S+M14	1	6.183	0.783	21.27
	2	27.14	0.6761	11.48
	4	31.97	0.6557	9.996
	6	44.29	0.5945	9.135
	8	52.22	0.551	8.794
Z5S+M16	1	7.036	0.7941	22.6
	2	20.76	0.6675	12.3
	4	31.81	0.6464	10.07
	6	85.05	0.5575	9.828
	8	99.37	0.5084	9.019

3.4.2. The Protective Performance Evaluated by Coating Resistance

The values of R_c, which can evaluate the protective performance of the coatings, are extracted from Table 7 and immersion time is plotted on the horizontal axis while R_c is plotted on the vertical axis (Figure 4). R_c decreases with the prolonging of immersion time for each content of micaceous iron oxide. In the earlier stage of immersion, the coating porosity increases due to the penetration of water as well as aggressive ions, while R_c is inversely proportional to coating porosity, so R_c decreases with immersion time. At the same immersion time, the coatings with higher micaceous iron oxide content have larger R_c. This may be due to the fact that the defects in the coating, as well as the coating porosity, gradually decrease as the micaceous iron oxide content gradually increase.

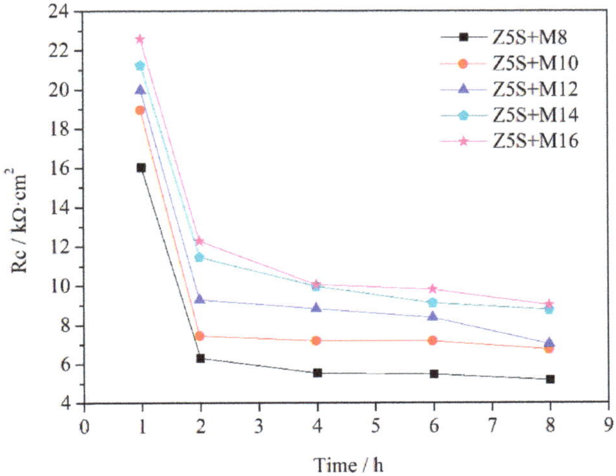

Figure 4. The changes of R_c curves as a function of immersion time.

3.4.3. The Protective Performance Evaluated by Breakpoint Frequency

Since breakpoint frequency f_* is related to the coating porosity or the coating defect area, the f_* can be used to evaluate the protective properties of the coating [51,52]. Cao [51] and Haruyama [53] provided the breakpoint frequency formula without considering the dispersion effect. Liu [39] used the −45° phase angle at a high frequency to obtain an improved breakpoint frequency formula considering the dispersion effect. However, some research [51] has shown that the phase angle corresponding to the f_* is no longer −45° for some coatings with the dispersion effect. It is similar to this investigation. Therefore, Liu's formula is not suitable for this paper. Haruyama [53] proposed an extreme value method to calculate the breakpoint frequency. This method is still available in the presence of the dispersion effect [51]. Accordingly, this method was adopted to deduce the breakpoint frequency formula.

Taking into account R_s and the influence of the dispersion effect, the expression of the impedance corresponding to the equivalent circuit R(QR) shown in Figure 3 is

$$Z = R_s + \frac{\frac{1}{R_c} + Y_0\omega^n \cos\left(\frac{n\pi}{2}\right) - jY_0\omega^n \sin\left(\frac{n\pi}{2}\right)}{\left(\frac{1}{R_c}\right)^2 + \left(\frac{2}{R_c}\right)Y_0\omega^n \cos\left(\frac{n\pi}{2}\right) + (Y_0\omega^n)^2} \tag{5}$$

where Y_0 is the constant of the constant-phase element (CPE), which is the value of Q_c in Table 7 and ω is the angular frequency.

Therefore, Z_{im} is

$$Z_{im} = \frac{Y_0\omega^n \sin\left(\frac{n\pi}{2}\right)}{\left(\frac{1}{R_c}\right)^2 + \left(\frac{2}{R_c}\right)Y_0\omega^n \cos\left(\frac{n\pi}{2}\right) + (Y_0\omega^n)^2} \tag{6}$$

The extreme value of Z_{im} should obey the following formula [53]:

$$\frac{dZ_{im}}{d\omega} = \frac{nY_0\omega^{n-1}\sin\left(\frac{n\pi}{2}\right)\left[\left(\frac{1}{R_c}\right)^2 + \left(\frac{2}{R_c}\right)Y_0\omega^n\cos\left(\frac{n\pi}{2}\right) + (Y_0\omega^n)^2\right] - \left[n\left(\frac{2}{R_c}\right)Y_0\omega^{n-1}\cos\left(\frac{n\pi}{2}\right) + 2(Y_0\omega^n)(nY_0\omega^{n-1})\right]\left[Y_0\omega^n\sin\left(\frac{n\pi}{2}\right)\right]}{\left[\left(\frac{1}{R_c}\right)^2 + \left(\frac{2}{R_c}\right)Y_0\omega^n\cos\left(\frac{n\pi}{2}\right) + (Y_0\omega^n)^2\right]^2} = 0 \tag{7}$$

That is

$$\omega_*{}^n = \frac{1}{R_C Y_0} \tag{8}$$

where ω_* is the characteristic angular frequency. The relationship between ω_* and f_* is

$$f_* = \frac{\omega_*}{2\pi} \tag{9}$$

The data in Table 7 are substituted into Equation (8) to obtain ω_* and then ω_* is substituted into Equation (9) to obtain f_*. The breakpoint frequency calculated by the above mentioned equations was plotted with immersion time in Figure 5.

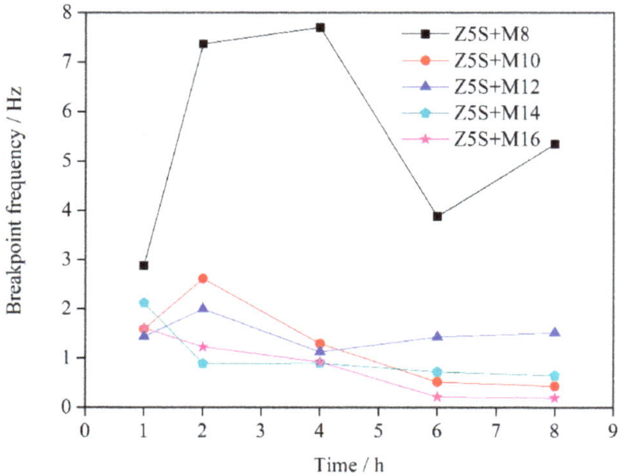

Figure 5. The changes of f_* curves as a function of immersion time.

The change range of f_* with immersion time is larger when the micaceous iron oxide content is 8%. This is mainly due to the fact that micaceous iron oxide has a disorderly arrangement in the coating when the contents of micaceous iron oxide are relatively lower, as shown in Figure 6a. It affects the quantity and distribution of defects in the coating to some extent, leading to a larger difference in the quantity of defects in all directions of the coating. The relationship between f_* and defect area (A_d) is $f_* = K\frac{A_d}{A}$, where $K \approx \frac{d}{(2\pi\varepsilon\varepsilon_0 r_t)}$, A is coating area, d is coating thickness, ε is the dielectric constant of coating containing corrosive media, ε_0 is the dielectric constant in vacuum, r_t is electrochemical reaction resistance per unit area [52]. The A and the d are constant in this experiment. The ε increases while the r_t decreases as corrosive media penetrate into the coating. This opposite effect can make the K be regarded approximately as a constant. So f_* is proportional to the A_d. Therefore, f_* shows sharp fluctuation with immersion time when the A_d changes drastically with immersion time due to lower micaceous iron oxide content. The fluctuation of f_* with immersion time is weakened as the contents of micaceous iron oxide increase. This is mainly due to the fact that with the increase of contents of micaceous iron oxide, its degree of disordered arrangement is reduced while the degree of ordered arrangement is increased (the micaceous iron oxide flakes are uniformly arranged in the coating and the flakes are alternately arranged in the cross-section direction of the coating, as shown in Figure 6b), which results in a reduction in the quantity of defects in the coating and an increase in the uniform of the distribution of defects in all directions of the coating. Therefore, the change range of the A_d with immersion time decreases. Since f_* is proportional to the A_d, the fluctuation of f_* decreases with immersion time. When the content of micaceous iron oxide reaches 16%, f_* is small and the fluctuation of f_* with immersion time is also small. It indicates that defects in the coating are few and their distribution is uniform as well as corrosive media diffusing slowly in the coating.

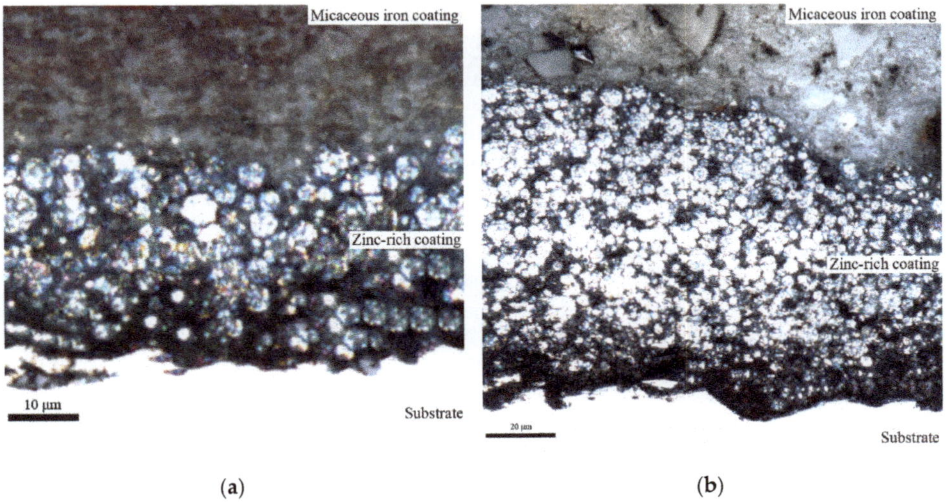

Figure 6. The CLSM images of the cross section of the coatings (**a**) Z5S + M10, (**b**) Z5S + M14.

When the contents of micaceous iron oxide are 10% to 12%, both f_* and A_d firstly increase, then decrease and finally tends to be stable as immersion time extends. This may be due to the fact that when the contents of micaceous iron oxide are relatively lower, micaceous iron's degree of ordered arrangement is not very high, which results in a certain amount of micaceous irons to be irregularly arranged, as shown in Figure 7. When corrosive media diffuse in the direction shown in Figure 7 as immersion time increases, A_d begins to increase gradually. As corrosive media continue to diffuse along the path shown in Figure 7, some of micaceous iron oxides may be slightly rotated or moved so that defects in their vicinity are filled or disappeared. After defects are reduced or minified, corrosive media are transferred to a "labyrinth structure" composed of regularly arranged micaceous iron oxides, where the coating has fewer defects and corrosive media can only diffuse along a tortuous path, resulting in a slow diffusion rate. Therefore, it appears that the A_d firstly increases, then decreases and finally tends to be stable.

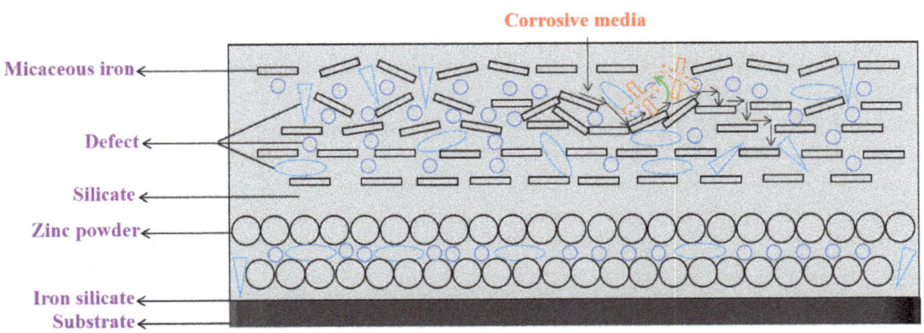

Figure 7. The corrosive media penetration model of low micaceous iron oxide contents.

When the contents of micaceous iron oxide are 14% to 16%, f_* trends initially decrease and subsequently become stable with immersion time. When the contents of micaceous iron oxide are relatively higher, micaceous iron's degree of ordered arrangement is higher but in the orderly arranged micaceous iron oxide structure there are still some micaceous iron oxides that are in the special position, as shown in Figure 8. When corrosive media diffuse in the direction shown in Figure 8 as immersion time increases, which may cause micaceous iron oxide in the special position in the ordered structure to be slightly rotated or moved so that defects in their vicinity are filled or disappeared. After defects are reduced or minified, corrosive media are transferred to the orderly arranged micaceous iron oxide structure, where there are fewer defects and corrosive media can only slowly diffuse along the "labyrinth" formed by micaceous iron oxide. Therefore, as a whole, the A_d shows the evolution trend of initial decrease and subsequent stability as immersion time is extended.

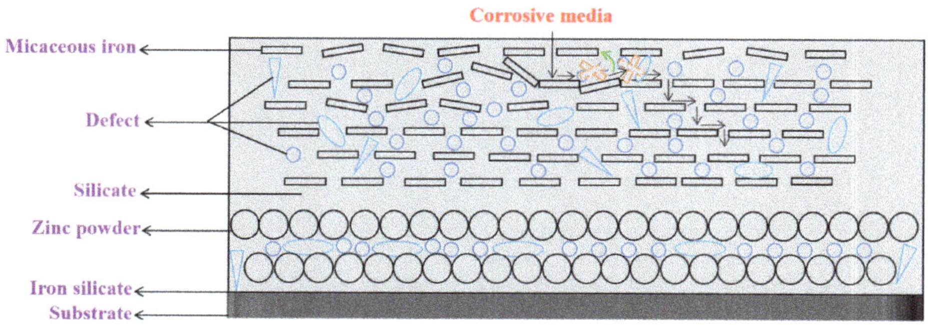

Figure 8. The corrosive media penetration model of high micaceous iron oxide contents.

Based on the above analysis and the Figure 5, it could be concluded that the f_* was less than 1.3 Hz from 2 h to 8 h when the micaceous iron oxide content exceeded 12%, indicating that the coating had a good protective structure.

4. Conclusions

In this paper, waterborne silicate zinc-rich antirust coatings, micaceous iron oxide anticorrosion coatings and composite coatings composed of antirust and anticorrosion ones were prepared and their properties were studied. For zinc-rich antirust coatings, when the particle size of zinc powder is constant, as the film thickness increases, the film forming effect becomes worse and the hydrophilicity of the coatings is enhanced. For micaceous iron oxide anticorrosion coatings, the adhesion and impact strength reached Grade 1 and 50 kg·cm, respectively, and the pencil hardness was up to 4H. These basic mechanical properties could meet the needs of the application on ships for anticorrosion coatings. Only one-time constant existed in the EIS results during the test period, which means that the composite coatings have good protection performance. The formula for breakpoint frequency, considering the dispersion effect, was deduced and used to evaluate the anticorrosive performance. It was found that the fluctuation of breakpoint frequency with immersion time was weakened as the content of micaceous iron oxide increased. Based on the relationship between breakpoint frequency and defect, different penetration models of corrosive media were proposed for the coatings with low or high contents of micaceous iron oxide.

Author Contributions: Conceptualization, X.Z., Y.Q. and Z.Z.; Methodology, X.Z., Y.Q. and Z.Z.; Software, Z.L.; Validation, K.L.; Formal Analysis, X.Z. and Z.Z.; Investigation, X.Z. and Z.Z.; Resources, Y.Q. and Z.Z.; Data Curation, X.Z.; Writing—Original Draft Preparation, X.Z.; Writing—Review and Editing, X.Z., Z.Z. and Y.Q.; Visualization, X.Z., K.L. and Z.L.; Supervision, Z.Z. and Y.Q.; Project Administration, Z.Z. and Y.Q.; Funding Acquisition, Z.Z. and Y.Q.

Funding: This research was funded by Project of Equipment Pre-research Field Fund, grant number 61409220304 and Equipment Pre-research Sharing Technology Project, grant number 41404010306 and 41423060314.

Conflicts of Interest: The authors declare no conflict of interest.

Appendix A

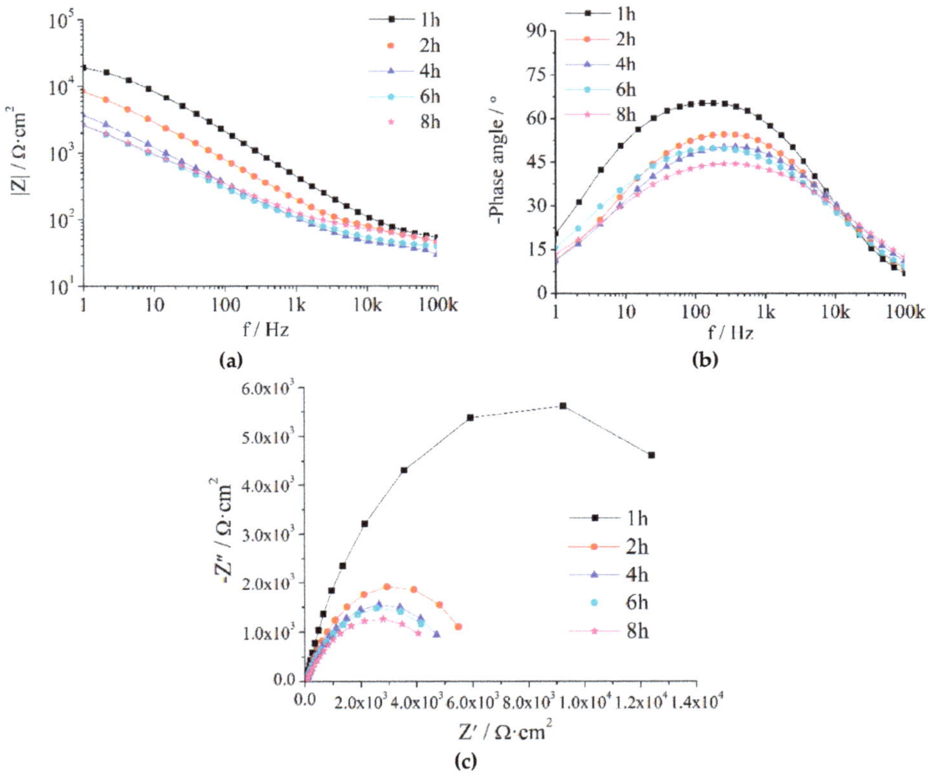

Figure A1. EIS of Z5S + M8 immersed in seawater for different times.

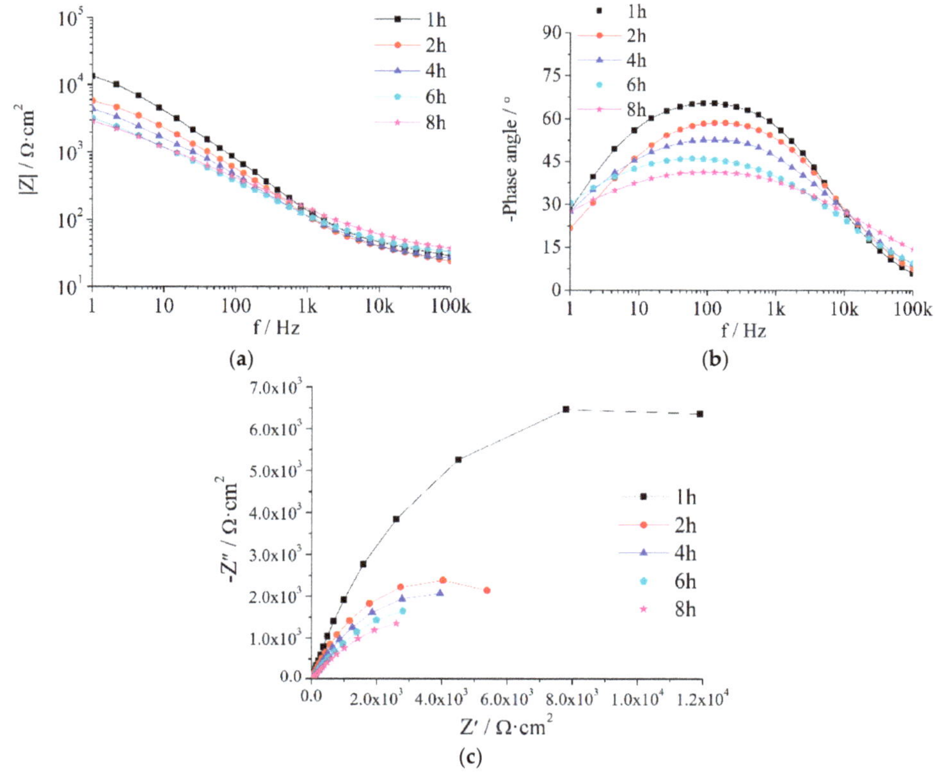

Figure A2. EIS of Z5S + M10 immersed in seawater for different times.

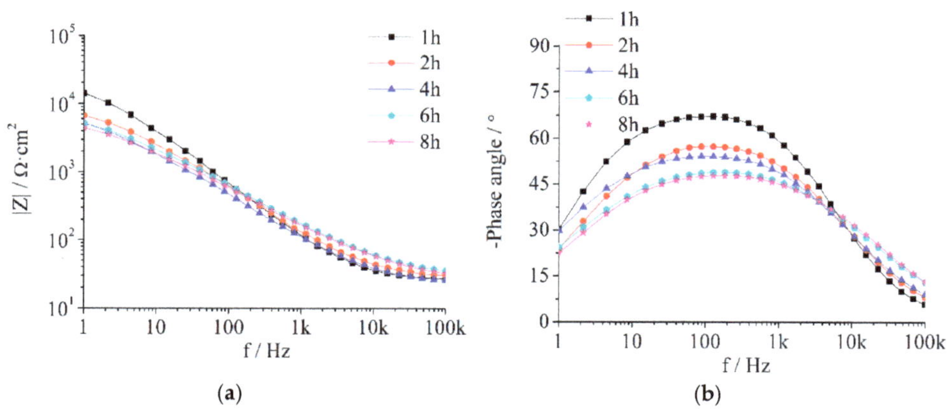

Figure A3. *Cont.*

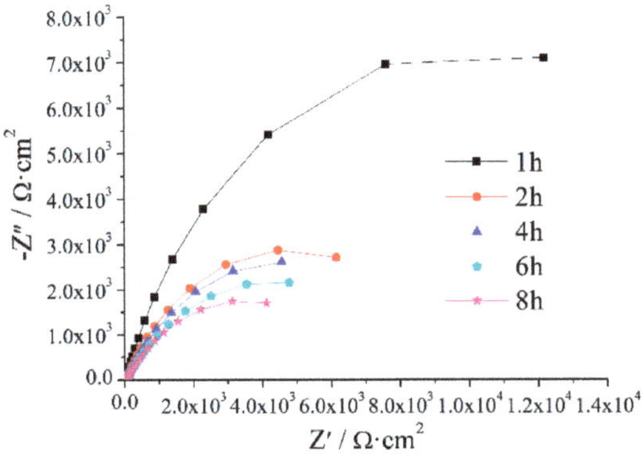

(c)

Figure A3. EIS of Z5S + M12 immersed in seawater for different times.

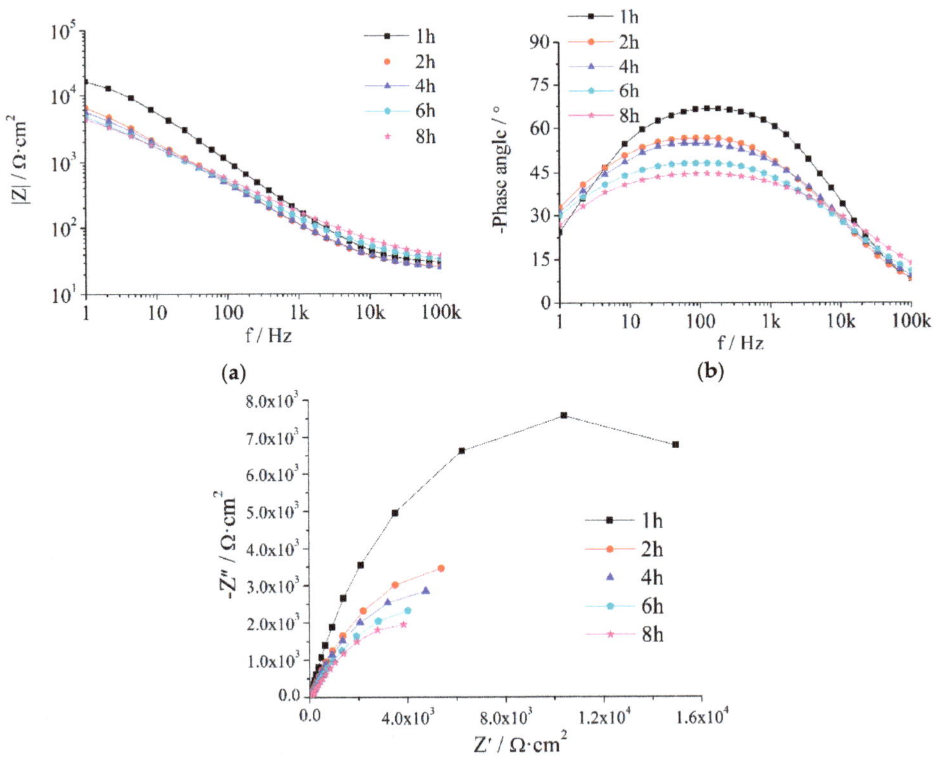

(c)

Figure A4. EIS of Z5S + M14 immersed in seawater for different times.

References

1. Wang, G.P. *Marine Coating and Coating Technology*, 2nd ed.; Chemical Industry Press: Beijing, China, 2006.
2. Wang, N.; Cheng, K.Q.; Zhang, J.; Zhang, D.; Kang, P. Method for preparing polyaniline modified mesoporous molecular sieve anticorrosion coating. CN Patent 201210109355.5, 22 August 2012.
3. Wang, N.; Wu, Y.H.; Cheng, K.Q.; Zhang, J. Investigation on anticorrosion performance of polyaniline-mesoporous MCM-41 composites in new water-based epoxy coating. *Mater. Corros.* **2015**, *65*, 968–976. [CrossRef]
4. Bagherzadeh, M.R.; Ghasemi, M.; Mahdavi, F.; Shariatpanahi, H. Investigation on anticorrosion performance of nano and micro polyaniline in new water-based epoxy coating. *Prog. Org. Coat.* **2011**, *72*, 348–352. [CrossRef]
5. Bagherzadeh, M.R.; Mahdavi, F.; Ghasemi, M.; Shariatpanahi, H.; Faridi, H.R. Using nanoemeraldine salt-polyaniline for preparation of a new anticorrosive water-based epoxy coating. *Prog. Org. Coat.* **2010**, *68*, 319–322. [CrossRef]
6. Wang, Z.; Han, E.; Liu, F.; Qian, Z.; Zhu, L. Waterborne epoxy nanocoatings modified by nanoemulsions and nanoparticles. *J. Mater. Sci. Technol.* **2014**, *30*, 1036–1042. [CrossRef]
7. Liu, J.H.; Zhan, Z.W.; L, S.M.; Yu, M. Corrosion resistance of waterborne epoxy coating pigmented by nano-sized aluminium powder on steel. *J. Cent. South Univ. Technol.* **2012**, *19*, 46–54. [CrossRef]
8. Jiang, C. A waterborne metal anticorrosion coating. CN Patent 201410488980.4, 25 February 2015.
9. Yao, Z.J.; Li, L.; Zhou, J.T.; Li, T.M.; Liu, P.J. Marine anticorrosion coating. CN Patent 201410337820.X, 8 October 2014.
10. Xu, J.; Rong, X.; Chi, T.; Ming, W.; Wang, Y.; Yang, D.; Qiu, F. Preparation, characterization of UV-Curable Waterborne Polyurethane-Acrylate and the application in metal iron surface protection. *J. Appl. Polym. Sci.* **2014**, *130*, 3142–3152. [CrossRef]
11. Ding, Y.S.; Ding, Y.Q.; Fang, H.G.; Yang, S.Z. An ultraviolet curing waterborne epoxy-polyurethane-acrylate copolymer resin-based anticorrosion paint and preparation method. CN Patent 201410275754.8, 10 September 2014.
12. Rashvand, M.; Ranjbar, Z. Effect of nano-ZnO particles on the corrosion resistance of polyurethane-based waterborne coatings immersed in sodium chloride solution via EIS technique. *Prog. Org. Coat.* **2013**, *76*, 1413–1417. [CrossRef]
13. Christopher, G.; Kulandainathan, M.A.; Harichandran, G. Highly dispersive waterborne polyurethane/ZnO nanocomposites for corrosion protection. *J. Coat. Technol. Res.* **2015**, *12*, 657–667. [CrossRef]
14. Rahman, O.U.; Kashif, M.; Ahmad, S. Nanoferrite dispersed waterborne epoxy-acrylate: Anticorrosive nanocomposite coatings. *Prog. Org. Coat.* **2015**, *80*, 77–86. [CrossRef]
15. Lin, L.; Xu, J. Study on Synthesis of nano-SiO_2/polyacrylate compound antirust emulsion. *Paint Coat. Ind.* **2014**, *44*, 19–22.
16. Lewis, O.D.; Critchlow, G.W.; Wilcox, G.D.; Dezeeuw, A.; Sander, J. A study of the corrosion resistance of a waterborne acrylic coating modified with nano-sized titanium dioxide. *Prog. Org. Coat.* **2012**, *73*, 88–94. [CrossRef]
17. Li, J.; Ecco, L.; Delmas, G.; Whitehouse, N.; Pan, J. In-Situ AFM and EIS Study of waterborne acrylic latex coatings for corrosion protection of carbon steel. *J. Electrochem. Soc.* **2015**, *162*, C55–C63. [CrossRef]
18. Sayyedan, F.S.; Enayati, M.H. On structure and oxidation behaviour of non-stoichiometric amorphous aluminium phosphate coating. *Surf. Eng.* **2019**, *35*, 670–676. [CrossRef]
19. Mokabber, T.; Zhou, Q.; Vakis, A.I.; van Rijn, P.; Pei, Y.T. Mechanical and biological properties of electrodeposited calcium phosphate coatings. *Mater. Sci. Eng. C Mater. Biol. Appl.* **2019**, *100*, 475–484. [CrossRef] [PubMed]
20. Castro, Y.; Duran, A. Control of degradation rate of Mg alloys using silica sol-gel coatings for biodegradable implant materials. *J. Sol-gel Sci. Technol.* **2019**, *90*, 198–208. [CrossRef]
21. Guo, L.L.; Tao, X.; Gong, Z.; Guo, A.R.; Du, H.Y.; Liu, J.C. Preparation of $MoSi_2$-SiC-Al_2O_3-SiO_2 coating on mullite fibrous insulation with silica sol as binder by non-firing process. *Ceram. Int.* **2019**, *45*, 2602–2611. [CrossRef]
22. Cheng, L.H.; Liu, C.L.; Han, D.J.; Ma, S.H.; Guo, W.H.; Cai, H.F.; Wang, X.H. Effect of graphene on corrosion resistance of waterborne inorganic zinc-rich coatings. *J. Alloy. Compd.* **2019**, *774*, 255–264. [CrossRef]

23. Jolin, W.C.; Oster, C.; Kaminski, M.D. Silicate coating to prevent leaching from radiolabeled surrogate far-field fallout in aqueous environments. *Chemosphere* **2019**, *222*, 106–113. [CrossRef]
24. Zhao, X.; Zhang, Z.Z.; Zhang, J.T.; Ding, G.Q. Research and development trend of environmentally friendly anticorrosive and antifouling coatings for hulls. *Shanghai Coat.* **2017**, *55*, 24–28. (In Chinese)
25. Giúdice, C.A.; Benítez, J.C. Optimising the corrosion protective abilities of lamellar micaceous iron oxide containing primers. *Anticorros. Methods Mater.* **2000**, *47*, 226–232. [CrossRef]
26. Sørensen, P.A.; Kiil, S.; Dam-Johansen, K.; Weinell, C.E. Anticorrosive coatings: A review. *J. Coat. Technol. Res.* **2009**, *6*, 135–176. [CrossRef]
27. Nikravesh, B.; Ramezanzadeh, B.; Sarabi, A.A.; Kasiriha, S.M. Evaluation of the corrosion resistance of an epoxy-polyamide coating containing different ratios of micaceous iron oxide/Al pigments. *Corros. Sci.* **2011**, *53*, 1592–1603. [CrossRef]
28. Wang, G.; Yang, J. Influences of glass flakes on fire protection and water resistance of waterborne intumescent fire resistive coating for steel structure. *Prog. Org. Coat.* **2011**, *70*, 150–156. [CrossRef]
29. Yang, C.; Smyrl, W.H.; Cussler, E.L. Flake alignment in composite coatings. *J. Membr. Sci.* **2004**, *231*, 1–12. [CrossRef]
30. Naser Kakaei, M.; Danaee, I.; Zaarei, D. Evaluation of cathodic protection behavior of waterborne inorganic zinc-rich silicates containing various contents of MIO pigments. *Anticorros. Methods Mater.* **2013**, *60*, 37–44. [CrossRef]
31. Castela, A.S.; Simões, A.M. An impedance model for the estimation of water absorption in organic coatings. Part II: A complex equation of mixture. *Corros. Sci.* **2003**, *45*, 1647–1660. [CrossRef]
32. Deflorian, F.; Fedrizzi, L.; Bonora, P.L. Influence of the photo-oxidative degradation on the water barrier and corrosion protection properties of polyester paints. *Corros. Sci.* **1996**, *38*, 1697–1708. [CrossRef]
33. Barranco, V.; Carpentier, J.; Grundmeier, G. Correlation of morphology and barrier properties of thin microwave plasma polymer films on metal substrate. *Electrochim. Acta* **2004**, *49*, 1999–2013. [CrossRef]
34. Brasher, D.M.; Kingsbury, A.H. Electrical measurements in the study of immersed paint coatings on metal. I. Comparison between capacitance and gravimetric methods of estimating water uptake. *J. Chem. Technol. Biotechnol.* **2010**, *4*, 62–72. [CrossRef]
35. Castela, A.S.; Simões, A.M. An impedance model for the estimation of water absorption in organic coatings. Part I: A linear dielectric mixture equation. *Corros. Sci.* **2003**, *45*, 1631–1646. [CrossRef]
36. Castela, A.S.; Simões, A.M. Assessment of water uptake in coil coatings by capacitance measurements. *Prog. Org. Coat.* **2003**, *46*, 55–61. [CrossRef]
37. Neshati, J.; Fardi, M.R. Evaluation and investigation of surface treatment of industrial coatings by impedance spectroscopy. *Surf. Eng.* **2013**, *20*, 299–303. [CrossRef]
38. Sykes, J.M. A variant of the Brasher–Kingsbury equation. *Corros. Sci.* **2004**, *46*, 515–517. [CrossRef]
39. Liu, L.; Hu, J.M.; Zhang, J.Q.; Cao, C.N. Evaluation of protectiveness of organic coatings by means of high-frequency EIS measurement. *Corros. Sci. Prot. Technol.* **2010**, *22*, 325–328.
40. Wang, P.; Yang, Q.; Cheng, X.R. Study on the curing mechanism of sodium silicate inorganic zinc-rich coating. *Appl. Chem. Ind.* **2007**, *36*, 1076–1080.
41. Peng, G.Y. Preparation and Properties of Waterborne Inorganic Coatings with Potassium Silicate/Silica Sol as Film-Forming Substance. Master's Thesis, South China University of Technology, Guangzhou, China, 2012.
42. Zhou, C.J. Preparation of Silicate Waterborne Inorganic Zinc-Rich Anticorrosion Coatings and Studies on Its Performance. Master's Thesis, Shenyang University of Technology, Shenyang, China, 2014.
43. *GB/T 1720-1979 Method of Test for Adhesion of Paint Films*; China Petroleum and Chemical Industry Federation: Beijing, China, 1980.
44. *GB/T 6739-2006 Paints and Varnishes-Determination of Film Hardness by Pencil Test*; China Petroleum and Chemical Industry Federation: Beijing, China, 2007.
45. *GB/T 1732-1993 Determination of Impact Resistance of Film*; China Petroleum and Chemical Industry Federation: Beijing, China, 1993.
46. Qi, Y.H.; Zhang, Z.P.; Miao, M.; Zhang, X.Z. Studies on estimating methods of polarization performance for coated steel in seawater. *Mater. Sci. Forum* **2010**, *654–656*, 2418–2421. [CrossRef]
47. Wenzel, R.N. Resistance of solid surfaces to wetting by water. *Ind. Eng. Chem.* **1936**, *28*, 988–994. [CrossRef]
48. Jiang, L.; Feng, L. *Bionic Intelligent Nano Interface Material*; Chemical Industry Press: Beijing, China, 2007.

49. Ding, J.H.; Liu, S.; Gu, L.; Zhao, H.C.; Yu, H.B. Corrosion resistance of epoxy phosphate/waterborne epoxy coatings on steel. *China Surf. Eng.* **2015**, *28*, 126–131. (In Chinese)
50. Li, Y.; Yang, Z.; Qiu, H.; Dai, Y.; Zheng, Q.; Li, J.; Yang, J. Self-aligned graphene as anticorrosive barrier in waterborne polyurethane composite coatings. *J. Mater. Chem. A* **2014**, *2*, 14139–14145. [CrossRef]
51. Cao, C.N.; Zhang, J.Q. *Introduction to Electrochemical Impedance Spectroscopy*; Science Press: Beijing, China, 2002.
52. Liu, H.; Liang, F.; Zhang, J.; Cao, C.; Shuqin, X.U.; Lin, H.; Dong, J.; Cai, Z. Investigation of effect of aluminium powder on performance of epoxy coatings by breakpoint frequency method. *Corros. Sci. Prot. Technol.* **1992**, *4*, 144–149.
53. Haruyama, S.; Asari, S.; Tsuru, T. Corrosion protection by organic coatings. *Electrochem. Soc.* **1987**, *87*, 197–207.

coatings

MDPI

Article

Characterisation of NiTi Orthodontic Archwires Surface after the Simulation of Mechanical Loading in CACO2-2 Cell Culture

Nikola Lepojević [1], Ivana Šćepan [1], Branislav Glišić [1], Monika Jenko [2], Matjaž Godec [2], Samo Hočevar [3] and Rebeka Rudolf [4,*]

1. Department of Orthodontics, School of Dental Medicine, University of Belgrade, Doktora Subotića 8, 11000 Belgrade, Serbia
2. Department of Surface Physics and Chemistry of Materials, Institute of Metals and Technology, Lepi pot 11, 1000 Ljubljana, Slovenia
3. Department of Analytical Chemistry, National Institute of Chemistry, Hajdrihova 19, 1001 Ljubljana, Slovenia
4. Faculty of Mechanical Engineering, University of Maribor, Smetanova 17, 2000 Maribor, Slovenia

* Correspondence: rebeka.rudolf@um.si

Received: 17 June 2019; Accepted: 10 July 2019; Published: 15 July 2019

Abstract: Nickel-titanium (NiTi) orthodontic archwires are crucial in the initial stages of orthodontic therapy when the movement of teeth and deflection of the archwire are the largest. Their great mechanical properties come with their main disadvantage—the leakage of nickel. Various in vitro studies measured nickel leakage from archwires that were only immersed in the medium with little or minimal simulation of all stress and deflection forces that affect them. This study aims to overcome that by simulating deflection forces that those archwires are exposed to inside the mouth of a patient. NiTi orthodontic archwires were immersed in CACO2-2 cell culture medium and then immediately loaded while using a simulator of multiaxial stress for 24 h. After the experiment, the surface of the NiTi orthodontic archwires were analysed while using scanning electron microscopy (SEM) and auger electron spectroscopy (AES). The observations showed significant microstructural and compositional changes within the first 51 nm thickness of the archwire surface. Furthermore, the released nickel and titanium concentrations in the CACO2-2 cell culture medium were measured while using Inductively Coupled Plasma Mass Spectroscopy (ICP-MS). It was found out that the level of released nickel ions was 1.310 μg/L, which can be assigned as statistically significant results. These data represent the first mention of the already detectable release of Ni ions after 24 h during the simulation of mechanical loading in the CACO2-2 cell culture medium, which is important for clinical orthodontic praxis.

Keywords: nickel-titanium; orthodontic archwires; surface; simulation; mechanical loading; CACO2-2 cell culture medium; scanning electron microscopy (SEM); auger electron spectroscopy (AES); inductively coupled plasma mass spectroscopy (ICP-MS)

1. Introduction

Today, shape memory alloys (SMA) have a wide variety of medical applications, including dentistry, especially for orthodontic treatment. They are attractive due to their superelasticity behaviour or functional property above the temperature austenite finish (A_f), and they are characterised by martensitic phase transformation, which is caused by the initiation of stress with enough value, leading to the changes of phase or microstructure in the SMA (austenite to martensite). Nickel-titanium (NiTi) orthodontic archwires are the most useful, where nickel and titanium are in equiatomic proportions—they are known as Nitinol. An ongoing challenge in the design of NiTi orthodontic archwires is to simultaneously have excellent biocompatibility, while also retaining adequate mechanical properties. The biocompatibility of NiTi orthodontic archwires has been investigated in many different studies [1–11]. Clinicians remain cautious of NiTi orthodontic archwire usage, because of their large nickel content, a long period of time being required for orthodontic treatment, and the high incidence of allergic reactions to nickel [1]. Several case reports describe allergic reactions of orthodontic patients to NiTi archwires [2,3]. One study concluded that there may be a risk of sensitising patients to nickel after long-term exposure [4]. Many studies proved the cytotoxicity of nickel, as it reduced the level of glutathione by binding to the sulfhydryl group of amino acids [5,6]. Nickel may also cause: lymphotoxicity [7], immunotoxicity [8], haemotoxicity [9], genotoxicity [10], and carcinogenicity [11]. Many studies have demonstrated that nickel release from these wires is very low and under the required threshold to cause biological effects, despite the high content of nickel in NiTi archwires [12–14]. The reason for this is that NiTi archwires have a high content of titanium, which makes it possible to form, with Ni, a stable crystal lattice with minimal potential release of nickel ions during the orthodontic treatment. On the other hand, titanium is a highly reactive element, which, once in contact with oxygen, immediately reacts to form a titanium oxide layer on the surface of the alloy. It is mainly composed of TiO_2 as the free enthalpy of formation of TiO_2 is negative, and it exceeds the enthalpy of the formation of nickel oxides by at least two- or three-fold in absolute value [15]. It protects the archwires from corrosion, and it creates a physical and mechanical barrier for the oxidation of nickel, leaving it deeper in the archwire [16]. This titanium oxide layer is very passive, but it could be removed or destroyed by many different factors in the mouth. Those archwires are under constant mechanical stress in the mouth, and they are simultaneously immersed in saliva, ingested fluids, temperature fluctuations, masticatory force, and topical fluoride modalities (toothpaste, mouthwashes, gels and varnish) [17], which also affect the surface of the archwires. Besides this, titanium has one great disadvantage, i.e., poor tribological properties and it exibits high friction and wear coefficients, as well as poor abrasion and fretting resistance when sliding against itself or other material [18]. Inside the mouth, the NiTi archwires are in constant contact with brackets that are made from different materials (stainless steel, ceramic, composite). Constant sliding and friction is inevitable in between them. Additionally, oxides, impurities, and pores are the key reasons for severe cavitation erosion damage of the NiTi [19].

Many in vitro studies examined NiTi archwires in static environments by putting them in different media and determining the level of released nickel [20,21]. Some experimental research deformed NiTi archwires only using one-dimensional loading, or failed to deform them at all. These procedures displayed results, whereby the level of released nickel showed little correspondence to the nickel that was released in the mouth. The Orthodontists deform NiTi archwire during the therapy and place it inside the bracket slot of every tooth (as shown in Figure 1a). The main stresses that deform NiTi archwires are the combination of bending stresses (tension and compressive) in conjunction with torsion and bending stresses. Archwires start unloading in order to come into a more stable austenitic phase and produce forces that move teeth after being placed inside the brackets. Teeth will move in a different manner, depending on the various levels of periodontal loss and the different stress distribution inside a healthy, endodontically treated, and restored tooth [22]. Fercec et al. [23] designed the simulation of multiaxial stress equipment (SMAS) that could provide uniaxial and multiaxial forces in order to deform the NiTi archwires (Figure 1b). With SMAS, different mechanical loading of the archwires can be simulated to imitate the real situation in the oral cavity. It is necessary to emphasize here that SMAS can only simulate intraoral forces that originate from the crowding of the teeth and the bending angle of the archwire, and cannot simulate frictional or masticatory forces. Based on the literature review, it was found that the experiment was not created to simulate contact of any fluid (medium) from the mouth with the archwires under loading. Consequently, the SMAS needed to be altered to facilitate contact of the NiTi archwires with the proper medium.

Based on this thesis, our reseach focused on the immersion of NiTi orthodontic archwires in medium and loading immediately, while using the SMAS for 24 h at an ambient temperature. The CACO2-2 cell culture medium was chosen, because it represents heterogeneous cells and it has found applications in cell invasion studies. The CACO-2 cell culture medium is used widely across the pharmaceutical industry as an in vitro model of the human small intestinal mucosa to predict the absorption of orally administered drugs with different kits [24]. Consequently, all of the metals that were released into the mouth of a patient during orthodontic treatment reach the intestines and CACO-2 cells as its crucial part. Various studies proved that nickel [25] and TiO_2 nanoparticles affect CACO-2 cells [26,27]. Those metals caused changes in cell viability, protein synthesis, geno-toxicity, oxidative stress, β-actin synthesis, and gene expression. All of those effects of nickel on cells were observed during 24 h, as in our study. The selected time, 24 h of simulation test, was chosen in accordance with the findings by Staffolani et al. [28], who discovered that nickel release from archwires reached its highest values on the first day. Some studies even point out that the initial Ni release increases are sustained, and they fail to drop over a prolonged period of a few months [29–32]. The amount of nickel that is released can vary, depending on the variable nickel surface concentrations that were reported for NiTi archwires (0.4–15 at.%) [33]. After the testing, demanding investigations were carried out, including the observation of changes on the NiTi orthodontic archwires' surface and measuring the ions' content in CACO2-2 cell culture medium as the results of release from the archwires. The logic of measuring the level of nickel in the CACO-2 cell line, instead of using the keratocytes, was that nickel released from the orthodontic archwire only stays a short amount of time in the mouth of a patient. It is ingested and, therefore, stays a much longer time in the colon, where it has higher chances to interact and affect cells. Similar, like in Pagano et al.' study [34], the cytotoxicity essay (MTT-3-(4,5-dimethylthiazol-2-yl)-2,5-diphenyltetrazolium bromide reduction assay) was planned to be done on CACO-2 cell culture, and to determine cell viability according to mitochondrial enzyme dehydrogenaze activity.

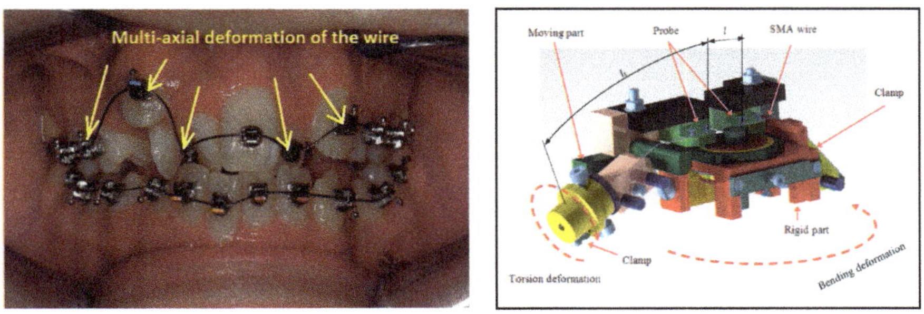

(a) (b)

Figure 1. (**a**) Fixed orthodontic appliance (arrow pointing to the bending and torsional stresses of the archwire); (**b**) Simulation of multiaxial stress equipment (SMAS) [23]. Reprinted with permission from [23]. © 2014 Elsevier.

The aim of this work was to dicover any changes in the surface of NiTi archwires and, if they are present, to measure the surface concentration of elements and released ion content into the medium during a 24 h deflection time. The null hypotheses of this study were: (1) There are no changes in the concentration of elements in the 51 nm of the wire surface, (2) A significant amount of nickel is not released from the wires into the medium, and (3) The amount of released nickel is the same for the deformed and initial archwires.

2. Materials and Methods

2.1. Materials

Commercially available NiTi archwires Rematitan (Dentaurum, Ispringen, Germany) dimensions 0.40 mm × 0.56 mm (0.016″ × 0.022″) were used in this study. The selected archwires had nearly equiatomic composition (50 at.% Ni, 50 at.% Ti). The American Type Culture Collection provided the

CACO-2 cell line (ATCC, Manassas, VA, USA) and it represented a continuous heterogeneous human epithelial colorectal adenocarcinoma cell line.

2.2. Methods

Mechanical loading in CACO2-2 cell culture SMAS was used for the simulation of NiTi orthodontic archwires [23] with the reconstruction, which represents the additional installation of a chamber (Figure 2a,b) with the volume 4 mL. The chamber was made from polymer material (plexiglass), which exhibits excellent mechanical and dimensional stability, and it is highly inert. It also lacks any metal that could contaminate the results, and it is not electro-conductive. The chamber construction was created, so that the CACO2-2 cell culture medium was in continuous contact with the archwires during testing in the SMAS: The archwires that were inside the SMAS went through the chamber in one part of their length. The chamber was closed with a cover that was made from the same material as the chamber to prevent leaking and evaporation of the liquid. Evaporation was prevented using the cover, which was further surrounded by silicone to entirely seal the chamber, therefore that amount of medium could be placed inside it during the experiment. For the purpose of this study, the archwires were immersed in CACO-2 cell culture medium inside the chamber.

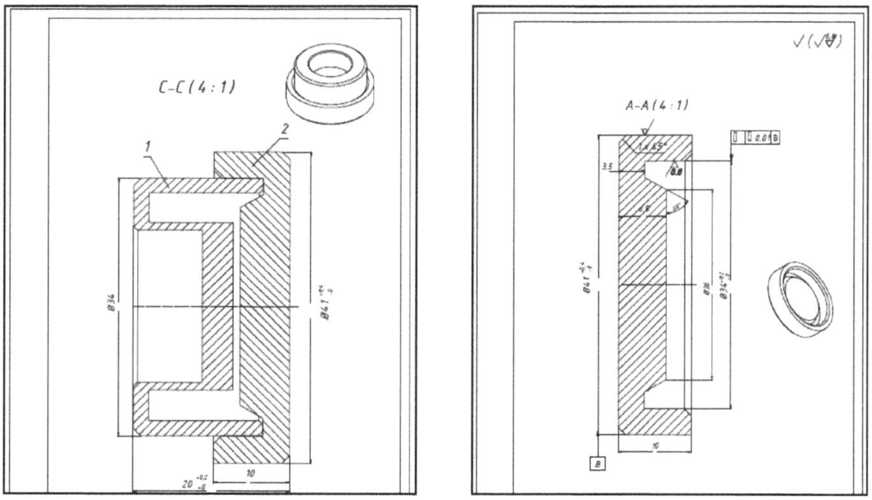

(**a**) (**b**)

Figure 2. Schematic presentation of (**a**) the chamber and (**b**) the cover.

2.3. Experimental Setup

In the simulation testing, orthodontic archwires Rematitan (Dentaurum, Ispringen, Germany) were placed in the SMAS (National Instruments OM 21 benchtop micro-ohmmeter, AOIP company, 52 Avenue Paul Langevin, 91130 Ris-Orangis, Paris, France) through the chamber and then bent at an angle of 30°, which represents the mean bending angle in the mouth of the average orthodontic patient. The length of the archwire in contact with CACO-2 cell culture medium was 3 cm. Four archwires were placed into the SMAS chamber at the same time in order to get as close to the situation of an average orthodontic patient, where about 12–15 cm length of orthodontic archwire is needed. After this, 4 mL of CACO-2 cell culture medium was added to the chamber and, in the next phase, the simulation test started for 24 h. A CLP load cell (HBM, Darmstadt, Germany) was used to measure the uni-axial tension force of the applied load. Displacement, which was caused by the screw, was measured from the initial distance to the flexible part.

2.4. Surface Analysis

The surfaces of NiTi orthodontic archwires after simulation testing were examined with scanning electron microscopy (SEM, Sirion 400 NC, FEI, Hillsboro, OR, USA) and auger electron spectroscopy (AES). The initial (non-deformed and non-immersed) orthodontic archwire served as a control. AES was performed while using a Microlab 310F VG-Scientific SEM/AES/X-ray photoelectron spectroscopy (XPS) (Thermo Scientific, Waltham, MA, USA), a field emission scanning spectrometer of auger electrons. A 10 kV primary electron beam was employed for this investigation, with a primary electron beam current of approximately 10 nA and approximate diameter of 10 nm resolution. The sample was Ar^+ sputtered with a sputter rate of 1 nm/min. for different sputter length times. AES spectrum was made after each sputtering time [35].

2.5. Measurement of Ions' Release

We prepared two types of testing in order to determine the impact of the load simulation on the release of ions from NiTi archwire in CACO-2 cell culture medium: (i) Exposure of NiTi archwires without any loading i.e., only immersed for 24 h (as a control sample) and (ii) Exposure with SMAS loading under the same conditions (four archwires, the same volume of medium, temperature, etc.). Inductively coupled plasma mass spectrometry (ICP-MS) while using a spectrometer (HP, Agilent 7500 ce, equipped with a collision cell, Santa Clara, CA, USA), was performed on both CACO-2 cell culture mediums to measure the concentration of the released nickel and titanium ions. ICP-MS analysis was carried out under the operating conditions: Power = 1.5 kW, Nebulizer = Meinhard, plasma gas flow (L/min) = 15, Nebulizer gas flow (L/min) = 0.85, Make up gas flow (L/min) = 0.28, and Reaction gas flow (mL/min) = 4.0. Prior to the analysis, the samples were dissolved in 10% (*v*/*v*) aqua regia. Matrix matched calibration solutions were prepared and analysed for calibration. Relative measurement uncertainty was established to be ±3%. The ICP-MS technique was used due to the advantages over other elemental analysers, as it is possible to perform quantitative analysis in a very large concentration range (from ng/mL in the sample solution until 100% in the sample). Besides this, correlation between the signal and concentration is linear for about five orders of magnitude for most of the analyses; there is a relatively good precision of measurements (around 3%), low matrix effects, and short analysis time.

3. Results and Discussion

3.1. Surface Analysis

The surface examination with SEM showed significant microstructure differences between the initial and deformed archwires (see Figure 3a,b). The initial archwire surface is rough, with visible holes and other defects, while the deformed archwire surface is smooth and without any other visible defects (inclusions, etc.). This can be attributed to the influence of mechanical loading, which led to the elongation and permanent deformation of the NiTi archwire [36] and the faster release of elements from the surface into the medium.

AES analyses were performed on the selected archwire surfaces (A1, A2—Figure 3a,b), the results are shown in Figure 4a,b. The AES detected a signal from the carbon, mostly originating from the C–C bonds at 280 eV, which suggested a residual contamination, probably from the atmosphere. Analysis of the spectra in the Ti 2*p* region indicates all the titanium to be present in the form of TiO_2. It has binding energies at 380 and 420 eV. The signal from the Ni 2*p* region was dominated by one large peak at 850 eV in both wires, but it is larger in the second deformed wire. There are also two smaller peaks at 730 and 780 eV that are similar in both of the wires. They correspond to the nickel oxides NiO and Ni_2O_3. The peak at 850 eV corresponds to elemental nickel. In Figure 4b, oxygen had one peak at approximately 530 eV, which pointed to a significant contribution from the Ti oxide, and upon closer inspection, another peak is observable at approximately 500 eV, which corresponds to Ni oxide. The deformed wire showed a higher level of nickel oxide at 500 eV despite the small amount of Ni oxide in the control wire.

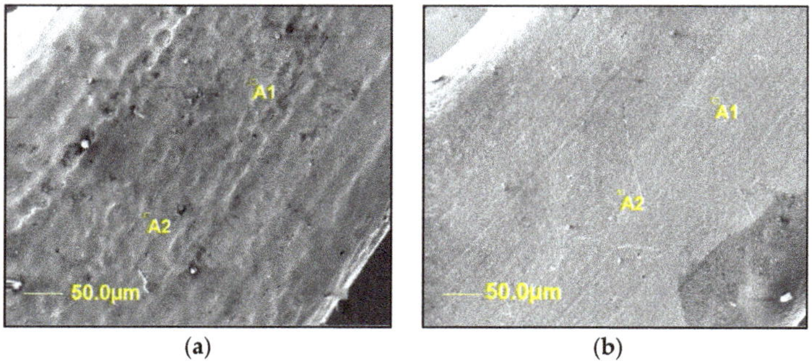

Figure 3. SEM image of: (**a**) Initial archwire surface (control); (**b**) Deformed archwire surface after the simulation of multiaxial stress equipment (SMAS) simulation test.

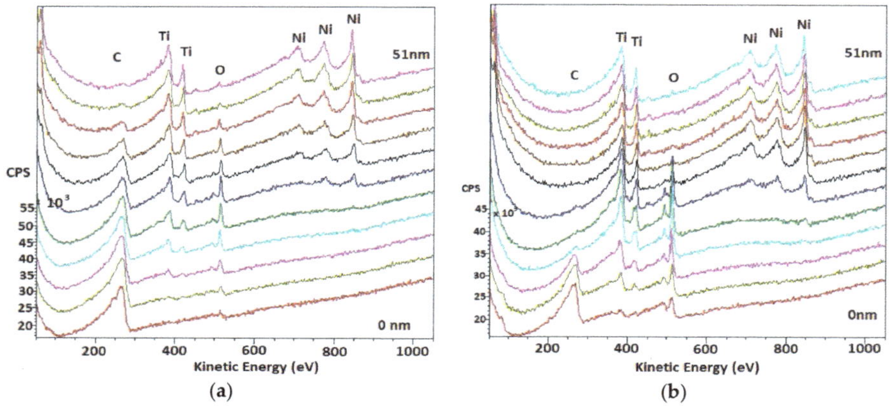

Figure 4. Auger electron spectroscopy (AES) spectra of archwire surfaces: (**a**) Initial (control); (**b**) Deformed after the SMAS simulation test.

Surface analysis AES showed concentration of elements in the first 51 nm of the selected archwires—Figure 5a,b depicts those results and discovered different concentrations of elements between the initial and deformed archwires. AES revealed the presence of carbon on the surface of both archwires. The binding energy of the carbon shows that it originates from the atmosphere (Figure 4a,b). It was detected that there is 40% less carbon present in the deformed archwire in 4 nm of depth, and an almost undetectable amount from 10 nm of depth, although, oxygen replaces carbon in the superficial layer of the wire in the deformed archwire (Figure 5b). Titanium is present on the surface of both archwires, and the nickel content is undetectable in the first 4 nm of both archwires. Therefore, it can be concluded that the oxygen is evidently connected to titanium, forming a protective layer of titanium oxides. In contrast to nickel, titanium with oxygen reacts immediately, and it forms a few oxides, but the most stable is in the form of TiO_2. The thickness of the oxide layer in the deformed archwire is about 51 nm, which is concluded from the fact that the level of oxygen drops to a concentration of 0% (Figure 5b). That is in accordance with the finding that the Ti-oxide film thickness ranges between 7 and 70 nm [37]. In contrast to this, the oxide layer is much thicker in the initial archwire, and the concentration of oxygen at a depth of 51 nm was still 12.2% (Figure 5a). The precise thickness of the oxide layer could not be detected on the initial archwire due to the fact that the AES was performed in the first 51 nm of the surface. Therefore, it could be concluded that the deformation of the archwire during 24 h affected the oxide thickness. The oxygen content of the deformed archwire differs from the initial archwire, and it is the highest at a depth of 6 nm. From Figure 4a,b it can be predicted that there is a significantly higher content of nickel oxide in the deformed archwire. The content of the nickel oxide is noticeable between 4 and 11 nm of the deformed archwire. From the results that were obtained, the amount of titanium is constantly increasing towards the bulk of the material in both archwires, and thus the amount of nickel shows a similar pattern in the initial archwire. However, this is slightly altered in the deformed archwire, where the rapid increase of nickel concentration can be noticed from the 11 nm depth. The difference of the nickel content is more than double at 16 nm depth between the archwires. From Figure 4b it can be concluded that most of the reason for that increased nickel content is in its elemental and oxidised state, which explains the phenomenon that, during the deformation of the archwire, nickel ions diffuse from the bulk of the material to the surface layer. Importantly, surface nickel ions dissolve more easily into the environment, similar to nickel ions during the heat treatment process that was described by Shabalovskaya et al. [33]. Besides oxide thickness, the surface nickel concentracion in NiTi shape memory alloys is a very important factor when considering the stability of the archwires [37]. The deformed archwire also exhibits a significant release of Ni ions due to the change in Ni/Ti ratio in contrast to the initial archwire. This is also followed by a thinner C-adsorption layer, which prevents the opening of the cracks in the Ti-oxide layer and the dissolution of Ni ions into the CACO-2 cell culture medium.

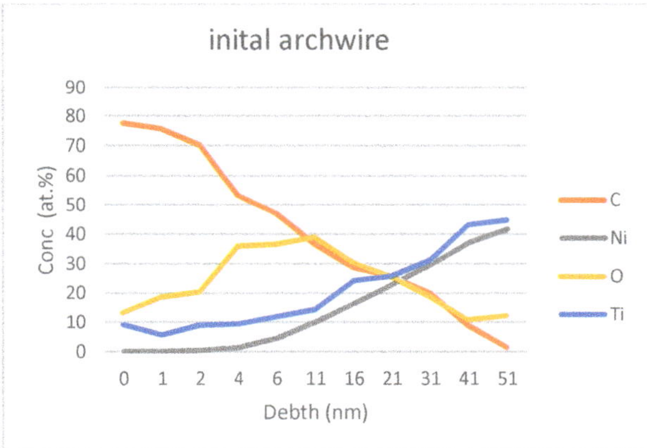

(**a**)

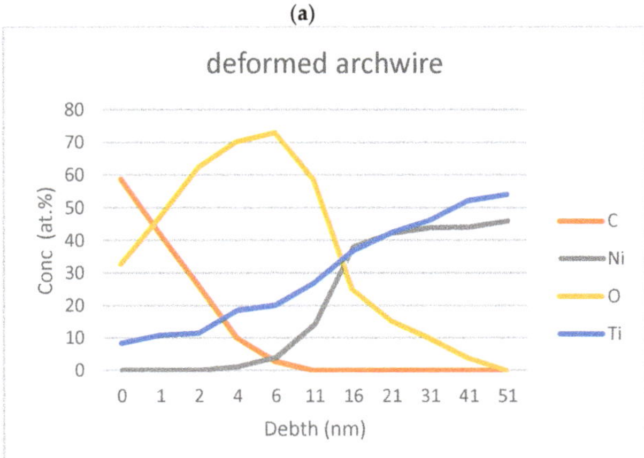

(**b**)

Figure 5. AES depth profiles show the concetrations of elements in the direction from the surface to 51 nm in the depth of each of the archwires: (**a**) Initial (control), (**b**) Deformed after the SMAS simulation test.

3.2. Results of Ions' Release

Table 1 shows the ICP-MS results of the ions' release in the CACO-2 cell culture mediums. The results indicated that titanium concentrations are below the detection limit (<0.05 μg/L) in both of the samples. On the other hand, the concentration of nickel is different; the difference was 70 μg/L/day. This suggested a small level of nickel leakage out of the archwire into the media.

Table 1. Ni and Ti ion concentrations from NiTi archwire in CACO-2 cell culture mediums after exposure with without loading (as a control sample) and after SMAS testing.

Sample	Conc Ni (μg/L)	Conc Ti (μg/L)
Control	1.240	<0.05
SMAS Testing	1.310	<0.05

Based on the obtained results, it can be concluded that the deformation of the archwire is an important factor that increases nickel release into the CACO-2 cell culture media. The concentration of nickel is 70 μg/L higher after 24 h in comparison with an unloaded archwire. Moreover, this represents the mean level of nickel released from a single wire of the average length of 12 cm (four pieces), while the patients usually have two arch wires in their mouth, and thus a total value of 140 μg/L additional nickel release. These results are in correlation with the literature, where similar levels of Ni release from similar archwires are mentioned during a comparable period of time [38,39]. The shape of the archwire can also influence the amount of nickel released, and rectangular archwires release more elements than round ones [40]. The value of released nickel is below the daily dietary intake level (300–500 μg) [41] and below the critical results (it must be taken into account that nickel can be released from a whole orthodontic appliance, which consists of bands, brackets, or some other attachments that also increase the nickel levels in the mouth). The physiological values are, therefore, much higher, and they can sometimes cause allergic reactions [42]. Furthermore, one study showed that even sub-toxic concentrations of metal ions can alter osteoblast activity [43]. Similarly, other authors [44] demonstrated that the 7.2 g/mL Ni ions released were sufficient for stimulating monocyte secretion of IL1, which, consequently, promoted endothelial cells to induce ICAM1 indirectly (Intracellular adhesion molecules that are involved in the activation of other Inflammatory cells). In addition, Cederbrant et al. [45] showed that an increase in lymphocyte proliferation and IL1 secretion could be induced, even with a small quantity of nickel, which is in correlation with our previous finding that thought archwires were non-cytotoxic for L929 cells, according to ISO Standards [15841:2014] [46], Rematitan superelastic archwires induced the apoptosis of rat thymocytes. This finding suggests strongly that, besides the released nickel ion concentration, the surface of the NiTi wire is most probably responsible for the cytotoxic effect [38]. It is important to declare that, in clinical conditions, orthodontic archwires are subjected to additional masticatory forces, abrasive forces from food and toothbrushes, and also temperature changes and various chemicals from saliva, liquids, and medicine. These can all damage the protective surface titanium oxide layer and cause corrosion, which leads to an increased level of released nickel ions. A few studies concluded that saliva, probiotic supplement and oral antiseptics affect both the general and localised corrosion of NiTi archwires, which also affects the mechanical properties and the release of nickel from NiTi archwires [47,48]. The oral environment is harsh and the conditions can change every second, so the archwires that are placed in the mouth need to resist all of those changes, including corrosion. Sometimes orthodontic treatment is complex [49,50] and the forces that are produced are much stronger than the ones with an average patient. Due to this complex situation, it is difficult to simultaneously simulate all of those conditions during in vitro studies.

It is necessary to improve the surface as a research challenge, and for safe use of NiTi archwires in the future. Namely, the presented study showed that the release of nickel ions from NiTi archwires is not negligible. According to the literature, it is known that one possible way could be nitriding of titanium with the formation of TiN/TiN_2, as nitrogen has a higher hardening effect than oxygen [51,52].

The other possible option could be the application of different coatings onto the NiTi surface that can improve its characteristics [53]. From other studies it was found out that there are good results in improving the frictional, biological, and aesthetic properties of stainless steel with electrophoretic deposition of a bioactive glass coating that could also be implemented on NiTi archwires [54].

4. Conclusions

Within the present study of NiTi orthodontic archwires' simulation loaded with multiaxial stresses in CACO2-2 cell culture, specific conclusions could be drawn.

The deformation of the NiTi archwire up to 30° during 24 h is enough to produce surface structural changes. SEM investigations of the deformed archwire surface revealed evident modification in the roughness, and the surface became smoother. On this surface, 40% less carbon was detected in comparison with the initial archwire surface. Besides this, the formatted thickness of the oxide layer on the surface was about 51 nm, while this could not be detected in the case of the initial archwire surface. It was found out that deformation accelerates the formation of oxides on the surface and with this oxide thickness, where a significantly higher content of nickel oxide was detected in the depth between 4 and 11 nm below the archwire surface. The oxygen content of the deformed archwire surface differed from the initial archwire surface, and it reached the highest value at a depth of 6 nm. This could be attributed to the phenomenon that, during the deformation of the archwire, nickel ions diffused faster from the bulk of the archwire surface.

Measurable nickel release during SMAS simulation of an archwire 12 cm in length is 1.310 μg/L, even after short period of 24 h.These data could be important for clinical orthodontic praxis. Additionally, the next studies must investigate the deformation time of the archwires, when considering the fact that NiTi orthodontic archwires could be in the oral environment for months, which can have considerable clinical implications. Based on the obtained results, it is recommended that, also in the case of biocompatibility, studies must be conducted to investigate the influence of archwire loading on the possible increase of nickel release content in comparison with the existing tests in order to obtain the most realistic information regarding surface stability.

Author Contributions: Conceptualization, R.R. and N.L.; Methodology, I.Š., B.G. and R.R.; Software, N.L., Validation, N.L., I.Š., B.G. and R.R.; Formal Analysis, M.J., M.G. and S.H.; Investigation, N.L., R.R., M.J., M.G. and S.H.; Writing—Original Draft Preparation, N.L. and R.R.; Writing—Review and Editing, R.R.; Funding Acquisition, I.Š., B.G., M.J. and R.R.

Funding: This research was funded by the Slovenian Research Agency ARRS (I0-0029–Infrastructure program University of Maribor) and by Eureka program ORTO-NITI E!6788 funded by Ministry of Education, Science and Technological Development Republic of Serbia.

Acknowledgments: The responsible proof reader for the English language is Shelagh Hedges, Faculty of Mechanical Engineering, University of Maribor, Slovenia.

Conflicts of Interest: The authors declare no conflict of interest.

References

1. Sidebottom, A.J.; Mistry, K. Prospective analysis of the incidence of metal alergy in patients listed for total replacement of the temporomandibular joint. *Br. J. Oral Maxillofac. Surg.* **2014**, *52*, 85–86. [CrossRef] [PubMed]
2. Dunlap, C.L.; Vincent, S.K.; Barker, B.F. Allergic reaction to orthodontic wire: Report of case. *J. Am. Dent. Assos.* **1989**, *118*, 449–450. [CrossRef] [PubMed]
3. Al-Waheidi, E.M.H. Allergic reaction to nickel orthodontic wires: A case report. *Quintessence Int.* **1995**, *26*, 385–387. [PubMed]
4. Bass, J.K.; Fine, H.; Cisneros, G.J. Nickel hypersensitivity in the orthodontic patient. *Am. J. Orthod. Dentofacial Orthop.* **1993**, *103*, 280–285. [CrossRef]
5. Dass, K.; Buchner, V. Effects of nickel exposure on peripheral tissues: role of oxidative stress in toxicity and possible protection by ascorbic acid. *Rev. Environ. Health* **2007**, *22*, 157–173. [CrossRef]

6. Valko, M.; Morriss, H.; Cronin, M.T. Metals toxicity and oxidative stress. *Curr. Med. Chem.* **2005**, *12*, 1161–1208. [CrossRef] [PubMed]
7. M'Bemba-Meka, P.; Lemieux, N.; Chakrabarti, S.K. Role of oxidative stress, mitochondrial membrane potential and calcium homeostasis in human lymphocyte death induced by nickel carbonate hydroxide in vivo. *Arch. Toxicol.* **2006**, *80*, 405–420. [CrossRef] [PubMed]
8. Dieter, M.P.; Jameson, C.W.; Tucker, A.N.; Luster, M.I.; French, J.E.; Hong, H.L.; Boorman, G.A. Evaluation of tissue deposition myelopoetic and immunologic responses in mice after long term exposure to nickel sulfare in drinking water. *J. Toxicol. Environ. Health* **1988**, *24*, 357–372. [CrossRef]
9. Hostynek, J.J. Sensitization to nickel: etiology, epidemiology, immune reactions, prevention and therapy. *Rev. Environ. Health* **2006**, *21*, 252–280. [CrossRef]
10. Liang, R.; Senturker, S.; Shi, X.; Bal, W.; Dizdaro-gluand, M.; Kasprzak, K.S. Effects of Ni (II) and Cu (II) on DNA interactions with the N-terminal sequence of human protamine P2: Enhancement of binding and mediation of oxidative DNA strand scission and base damage. *Carcinogenesis* **1999**, *20*, 893–898. [CrossRef]
11. Kasprzak, K.S.; Sunderman, F.W.J.; Salnikow, K. Nickel carcinogenesis. *Mutat. Res.* **2003**, *533*, 67–97. [CrossRef] [PubMed]
12. Rahilly, G. Nickel allergy and orthodontics. *J. Orthod.* **2003**, *30*, 171–174. [CrossRef] [PubMed]
13. Huang, H.H.; Chiu, Y.H.; Lee, T.H.; Wu, S.C.; Yang, H.W.; Su, K.H.; Hsu, C.C. Ion release from NiTi orthodontic wires in artificial saliva with various acidities. *Biomaterials* **2003**, *24*, 3585–3592. [CrossRef]
14. Ryhanen, J.; Niemi, E.; Serlo, W.; Niemela, E.; Sandvik, P.; Pernu, H.; Salo, T. Biocompatibility of nickel-titanium shape memory metal and its corrosion behavior in human cell cultures. *J. Biomed. Mater. Res.* **1997**, *35*, 451–457. [CrossRef]
15. Chan, C.; Trigwell, C.; Duerig, T. Oxidation of a NiTi alloy. *Surf. Interface Anal.* **1990**, *15*, 349–354. [CrossRef]
16. Espinos, J.P.; Fernandes, A.; Gonzales-Elipe, A.R. Oxidation and diffusion proccesses in nickel-titanium oxide systems. *Surf. Sci.* **1993**, *295*, 402–410. [CrossRef]
17. Heravi, F.; Moayed, M.H.; Mokhber, N. Effects of fluoride on nickel-titanium and stainless-steel orthodontic archwires: An in vitro study. *J. Dent.* **2015**, *12*, 49–59.
18. Qu, J.; Blau, P.J.; Watkins, T.R.; Cavin, O.B.; Kulkarni, N.S. Friction and wear of titanium alloys sliding against metal, polymer and ceramic counterfaces. *Wear* **2005**, *258*, 1348–1356. [CrossRef]
19. Zhenping, S.; Jiqiang, W.; Zhengbin, W.; Yanxin, Q.; Tianying, X.; Yugui, Z. Cavitation erosion and jet impingement erosion of the NiTi coating produced by air plasma spraying. *Coatings* **2018**, *8*, 346.
20. Freiberg, K.E.; Bremer-Streck, S.; Kiehntopf, M.; Rettenmayr, M.; Undisz, A. Effect of thermomechanical pre-treatment on short- and long-term Ni release from biomedical NiTi. *Acta Biomater.* **2014**, *10*, 2290–2295. [CrossRef]
21. Jia, W.; Beatty, M.W.; Reinhardt, R.A.; Petro, T.M.; Cohen, D.M.; Maze, C.R.; Strom, E.A.; Hoffman, M. Nickel release from orthodontic arch wires and cellular immune response to various nickel concentrations. *J. Biomed. Mater. Res.* **1999**, *48*, 488–495. [CrossRef]
22. Chieruzzi, M.; Pagano, S.; Cianetti, S.; Lombardo, G.; Kenny, J.M.; Torre, L. Effect of fibre post, bone losses and fibre content on the biomechanical behaviour of endodontically treated teeth: 3D-finite alement analysis. *Mater. Sci. Eng. C* **2017**, *74*, 334–346. [CrossRef] [PubMed]
23. Fercec, J.; Anzel, I.; Rudolf, R. Stress dependent electrical resistivity of orthodontic wire from the shape memory alloy NiTi. *Mater. Des.* **2014**, *55*, 699–706. [CrossRef]
24. Vázquez-Sánchez; Ángeles, M. Available online: https://readycell.com/cacoready/ (accessed on 19 July 2018).
25. Calabro, A.R.; Gazarian, D.I.; Barile, F.A. Effect of metals on β-actin and total protein synthesis in cultured human intestinal epithelial cells. *J. Pharmacol. Toxicol. Methods* **2011**, *63*, 47–58. [CrossRef] [PubMed]
26. Richter, J.W.; Shull, G.M.; Fountain, J.H.; Guo, Z.; Musselman, L.P.; Fiumera, A.C.; Mahler, G.J. Titanium dioxide nanoparticle exposure alters metabolic homeostasis in a cell culture model of the intestinal epithelium and drosophila melanogaster. *Nanotoxicology* **2018**, *12*, 390–406. [CrossRef] [PubMed]
27. García-Rodríguez, A.; Vila, L.; Cortés, C.; Hernández, A.; Marcos, R. Effects of differently shaped TiO_2NPs (nanospheres, nanorods and nanowires) on the in vitro model (Caco-2/HT29) of the intestinal barrier. *Part. Fibre Toxicol.* **2018**, *15*, 33. [CrossRef] [PubMed]
28. Staffolani, N.; Damiani, F.; Lilli, C.; Guerra, M.; Staffolani, N.J.; Belcastro, S.; Locci, P. Ion release from orthodontic appliances. *J. Dent.* **1999**, *27*, 449–454. [CrossRef]

29. Cisse, O.; Savagodo, O.; Wu, M.; Yahia, L. Effect of surface treatment of NiTi alloy on its corrosion behavior in Hanks solution. *J. Biomed. Mater. Res.* **2002**, *61*, 339–345. [CrossRef]
30. Kobayashi, S.; Ohgoe, Y.; Ozeki, K.; Sato, K.; Sumiya, T.; Hirakuri, K. Diamond-like carbon coatings on orthodontic archwires. *Diam. Relat. Mater.* **2005**, *14*, 1094–1097. [CrossRef]
31. Sui, J.; Cai, W. Effect of diamond-like carbon (DLC) on the properties of NiTi alloys. *Diam. Relat. Mater.* **2006**, *15*, 1720–1726. [CrossRef]
32. Clarke, B.; Carroll, W.; Rochev, Y.; Hynes, M.; Bradley, D.; Plumley, D. Influence of nitinol wire surface treatment on oxide thickness and composition and its subsequent effect on corrosion resistance and nickel ion release. *J. Biomed. Mater. Res. A* **2006**, *79*, 61–70. [CrossRef] [PubMed]
33. Shabalovskaya, S.; Anderegg, J.; Laab, F.; Thiel, P.A.; Rondelli, G. Surface conditions of Nitinol wires, tubing, and as-cast alloys. The effect of chemical etching, aging in boiling water, and heat treatment. *J. Biomed. Mater. Res.* **2003**, *65B*, 193–203. [CrossRef] [PubMed]
34. Pagano, S.; Chieruzzi, M.; Balloni, S.; Lombardo, G.; Torre, L.; Bodo, M.; Cianetti, S.; Mairnucci, L. Biological, thermal and mechanical characterization of modified dlass ionomer cements: The role of nanohydroxyapatite, ciprofloxacin and zinc L-carnosine. *Mater. Sci. Eng. C* **2019**, *94*, 76–85. [CrossRef] [PubMed]
35. Shabalovskaya, S.; Anderegg, J. Surface spectroscopic characterization of TiNi nearly equiatomic shape memory alloys for implants. *J. Vac. Sci. Technol. A* **1995**, *13*, 2624–2632. [CrossRef]
36. Uchil, J.; Mahesh, K.K.; Ganesh-Kumara, K. Electrical resistivity and strain recovery studies on the effect of thermal cycling under constant stress on R-phase in NiTi shape memory alloy. *Phys. B Condens. Mater.* **2002**, *324*, 419–428. [CrossRef]
37. Shabalovskaya, S.A. On the nature of the biocompatibility an medical applications of NiTi shape memory and superelastic alloys. *Biomed. Mater. Eng.* **1996**, *6*, 267–289. [PubMed]
38. Colic, M.; Tomic, S.; Rudolf, R.; Markovic, E.; Scepan, I. Differeces in citocompatibility, dynamics of the oxide layers' formation, and nickel release between superelastic and thermo-elastic activated nickel-titanium archwires. *J. Mater. Sci. Mater. Med.* **2016**, *27*, 128.
39. Ramazanzadeh, B.A.; Ahrari, F.; Sabzevari, B.; Habibi, S. Nickel ion release from three types of nickel-titanium-based orthodontic archwires in the as-received state and after oral simulation. *J. Dent. Res. Dent. Clin. Dent. Prospects* **2014**, *8*, 71–76.
40. Azizi, A.; Jamilian, A.; Nucci, F.; Kamali, Z.; Hosseinikhoo, N.; Perillo, L. Release of metal ions from round and rectangular NiTi wires. *Prog. Orthod.* **2016**, *17*, 10. [CrossRef]
41. Schroeder, H.A.; Balassa, J.J.; Tipton, I.H. Abnormal trace metals in man-nickel. *J. Chron. Dis.* **1962**, *15*, 51–65. [CrossRef]
42. Kaaber, K.; Veien, N.K.; Tjell, J.C. Low nickel diet in the treatment of patients with chronic nickel dermatitis. *Br. J. Dermatol.* **1978**, *98*, 197–201. [CrossRef] [PubMed]
43. Sun, Z.L.; Wataha, C.T.; Hanks, C.T. Effects of metal ions on osteoblast-like cell metabolism and differentiation. *J. Biomed. Mater. Res.* **1997**, *34*, 29–37. [CrossRef]
44. Wataha, J.C.; Lockwood, P.E.; Marek, M.; Ghazi, M. Ability of Ni-containing biomedical alloys to activate monocytes and endothelial cells in vitro. *J. Biomed. Mater. Res.* **1999**, *45*, 251–257. [CrossRef]
45. Cederbrant, K.; Anderson, C.; Andersson, T.; Marcusson-Stahl, M.; Hultman, P. Cytokine production, lymphocyte proliferation and T-cell receptor Vβ expression in primary peripheral blood mononuclear cell cultures from nickel-allergic individuals. *Int. Arch. Allergy Immunol.* **2003**, *132*, 373–379. [CrossRef] [PubMed]
46. *ISO 15841:2014: Dentistry—Wires for Use in Orthodontics*; International Organization for Standardization: Geneva, Switzerland, 2014.
47. Rincic Mlinaric, M.; Karlovic, S.; Ciganj, Z.; Acev, D.P.; Pavlic, A.; Spalj, S. Oral antiseptics and nickel-titanium alloys: Mechanical and chemical effects of interaction. *Odontology* **2019**, *107*, 150–157. [CrossRef] [PubMed]
48. Trolić, I.M.; Turco, G.; Contardo, L.; Serdarević, N.L.; Ćurković, H.O.; Špalj, S. Corrosion of nickel-titanium orthodontic archwires in saliva and oral probiotic supplements. *Acta Stomatol. Croat.* **2017**, *51*, 316–325.
49. D'Attilio, M.; Rodolfino, D.; Filippakos, A.; Saccucci, M.; Festa, F.; Tripodi, D. Second class resolver: A retrospective analysis. *Eur. J. Paediatr. Dent.* **2014**, *15*, 78–82.
50. Baldini, A.; Nota, A.; Santariello, C.; Assi, V.; Ballanti, F.; Gatto, R.; Cozza, P. Sagittal dentoskeletal modifications associated with different activation protocols of rapid maxillary expansion. *Eur. J. Paediatr. Dent.* **2018**, *19*, 151–155.

51. Kamat, A.M.; Copley, S.M.; Segal, A.E.; Todd, J.A. Laser-sustained plasma (LSP) nitriding of titanium: A review. *Coatings* **2019**, *9*, 283. [CrossRef]
52. Al Jabbari, Y.S.; Fehrman, J.; Barnes, A.C.; Zapf, A.M.; Zinelis, S.; Berzins, D.W. Titanium nitride and nitrogen ion implanted coated dental materials. *Coatings* **2012**, *2*, 160–178. [CrossRef]
53. Arango, S.; Peláez-Vargas, A.; García, C. Coating and surface treatments on orthodontic metallic materials. *Coatings* **2013**, *3*, 1–15. [CrossRef]
54. Kawaguchi, K.; Iijima, M.; Endo, K.; Mizoguchi, I. Electrophoretic deposition as a new bioactive glass coating process for orthodontic stainless steel. *Coatings* **2017**, *7*, 199. [CrossRef]

coatings

MDPI

Article

Dual Component Polymeric Epoxy-Polyaminoamide Based Zinc Phosphate Anticorrosive Formulation for 15CDV6 Steel

Omar Dagdag [1], Ghadir Hanbali [2], Bayan Khalaf [2], Shehdeh Jodeh [2,*], Ahmed El Harfi [1] and Abdelhadi Deghles [3]

1 Laboratory of Agroresources, Polymers and Process Engineering (LAPPE), Department of Chemistry, Faculty of Science, Ibn Tofail University, BP 133, 14000 Kenitra, Morocco
2 Department of Chemistry, An-Najah National University, P.O. Box 7 Nablus, Palestine
3 Scientific Research and Development, Al Istiqlal University, P.O. Box 10 Jericho, Palestine
* Correspondence: sjodeh@hotmail.com; Tel.: +970-599590498

Received: 20 June 2019; Accepted: 12 July 2019; Published: 24 July 2019

Abstract: The present research is focused on a formulation with two active components as an anticorrosive polymer coating for 15CDV6 steel. The dual component formulation (epoxy-zinc phosphate (ZP) coating) consists of a polymeric epoxy resin Bisphenol A diglycidyl ether (DGEBA) cured with a polyaminoamide as a first component and zinc phosphate ($Zn_3(PO_4)_2(H_2O)_4$) (ZP) added in 5% by weight as a second component. The anticorrosive performance of the epoxy-ZP coating was evaluated against the standard coating, which consists of only one component, the cured polymeric epoxy resin. The two polymer coatings were evaluated by electrochemical impedance spectroscopy (EIS). The surface morphology was of the two polymer coatings were characterized by scanning electron microscopy (SEM). The coated samples of 15CDV6 steel were tested in a harsh environment of corrosive electrolytes (3 wt % NaCl solution). Under these conditions, a very high impedance value was obtained for 15CDV6 steel coated with the epoxy-ZP coating. Even after exposure for a long period of time (5856 h), the performance was still acceptable, indicating that the epoxy-ZP coating is an excellent barrier. The standard epoxy coating provided an adequate corrosion protection performance for a short period of time, then the performance started to decline. The results were confirmed by surface characterization, a cross-sectional image obtained by optical microscopy for an epoxy-ZP coating applied on 15CDV6 steel exposed for 5856 h to a salt spray test showed that the coating is homogeneous and adheres well to the surface of the steel. So, the coating with a dual component could have great potential in marine applications as anticorrosive for steel.

Keywords: polymeric epoxy resin; polyaminoamide; zinc phosphate; polymer coating; steel and saline

---

1. Introduction

Corrosion of steel alloys is one of the most important safety and economic concerns for many industries [1,2]. Because of its excellent mechanical power and relatively low cost, 15CDV6 steel is one of the frequently used steel-based materials for numerous applications in several industries. However, it is highly susceptible to corrosion during several industrial processes where metallic components undergo corrosive dissolution by the aggressive saline environment [3,4].

Organic coatings are usually applied as thin films on the surface of metals, thus preventing corrosive agents from reaching the metal surface [5–8]. Despite all the advantages of organic coatings, some improvement is still needed, for instance their life-time is limited, after a certain period they tend to deteriorate and their performance, such as their barrier properties, declines [9]. Durability and life time of an organic coating are controlled by several factors, among these are the chemistry of the coating, the cross-linking density in the coating, and the functional group present on the coating surface which affects the strength of adhesion to the metal surface. Among organic coatings, water-based epoxy coatings are the most attractive due to their unique mechanical properties, superior adhesion to substrate, good thermal stability, excellent corrosion resistance and chemical resistance. In addition, they can be available at a low cost [10,11] and are nontoxic. These attractive properties could be attributed to high the cross-linking density and to the functional groups such as amines and hydroxyl present in the polymer coating [12]. However, long term exposure to corrosive materials, cause partial diffusion of corrosive species into the polymer coating from scratches or inherent pores. Eventually the diffusing materials will reach the metal–coating interface. This causes the initiation of corrosion reactions and slow pealing of the coatings [13–15].

Using inorganic pigments such as zinc chromate as a corrosion inhibitor has become popular in recent years [16]. They tend to improve the adhesion and anticorrosion performance of organic coatings. A couple of published studies showed that some of the inorganic pigments dissolve in water. The dissolved part tends to undergo oxidation and precipitate as a passive film on the metal surface [17,18]. The passive film can block the active zones on the metal surface and reduce the rate of electrochemical processes [19].

However, the toxicity and the carcinogenicity of some of the reported inorganic pigment have limited their use [13,20,21]. In recent decades, alternative nontoxic pigments were developed such as, zinc phosphate (ZP). In addition, zinc phosphate pigment showed better performance as a corrosion inhibitive for steel. This was attributed to it is solubility in water.

In the first part of the present study, a polymeric based Bisphenol A diglycidyl ether (DGEBA) cured with a polyaminoamide was prepared and coated on the surface of 15CDV6 steel with zinc phosphate pigments.

In the second part, we have evaluated and tested in an aggressive marine environment. The anticorrosive performance of the two polymer coatings was monitored by electrochemical impedance spectroscopy (EIS) and confirmed by SEM.

2. Experimental

2.1. Materials and Methods

The formulations of the polymer coatings are given in Table 1.

Table 1. Composition of organic model coatings.

Component (Liquid Epoxy)	Weight Percent (wt %)	Supplier
DGEBA (average molecular weight < 700)	$25 \leq$ wt % < 50	MAPAERO-Aerospace Coatings and Finishes, Pamiers CEDEX, France
Nitroethane	$25 \leq$ wt % < 50	MAPAERO-Aerospace Coatings and Finishes, Pamiers CEDEX, France
Triglycidyl ether of trimethylolpropane	$10 \leq$ wt % < 25	MAPAERO-Aerospace Coatings and Finishes, Pamiers CEDEX, France
Alkoxysilane	$0 \leq$ wt % < 2.5	MAPAERO-Aerospace Coatings and Finishes, Pamiers CEDEX, France
Component (Liquid Hardener)	**Weight Percent (wt %)**	**Supplier**
Polyaminoamide	$10 \leq$ wt % < 25	MAPAERO-Aerospace Coatings and Finishes, Pamiers CEDEX, France
Butane-2-ol	$25 \leq$ wt % < 50	MAPAERO-Aerospace Coatings and Finishes, Pamiers CEDEX, France
Zinc oxide	$0 \leq$ wt % < 2.5	MAPAERO-Aerospace Coatings and Finishes, Pamiers CEDEX, France
Demineralized water ($\varrho > 1$ Ω·cm and $\sigma < 1$ μS/cm)	–	–
Methyl ethyl ketone (MEK) (≥99%)	–	Sigma Aldrich, Saint Louis, MO, USA
Zinc phosphate (ZP) (99.99%)	$0 \leq$ wt % < 5	Sigma Aldrich, Saint Louis, MO, USA

Samples of 15CDV6 steel were used in this study, the composition of the steel samples are given in Table 2.

Table 2. Elementary chemical composition in weight percent (wt %) of the 15CDV6 steel that was used.

Elements	wt % Composition
C	0.12–0.18
Si	≤0.2
Mn	0.8–1.1
S	≤0.015
P	≤0.02
Cr	1.25–1.5
Mo	0.8–1.0
V	0.2–0.3
Fe	Balance

2.2. *Preparation of 15CDV6 Steel Panels*

Samples of 15CDV6 steel were cut as rectangular panels of dimensions 120 cm × 40 cm × 2 cm. The exposed surface was polished using levigated alumina paste as abrasive in order to obtain mirror surface then degreased with Methyl ethyl ketone (MEK), cleaned by demineralized water, then dried.

2.3. *Epoxy Resin Formulation and Steel Coating*

Two formulations were prepared, one of them had a hardener in addition to the DGEBA (1:2 ratio by weight) (standard epoxy coating) (Figure 1).

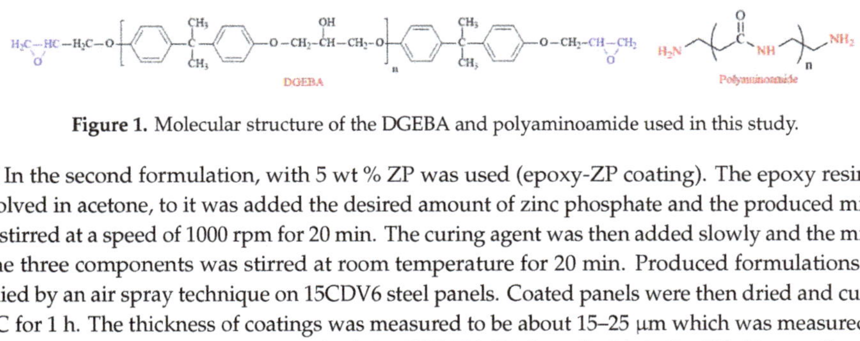

Figure 1. Molecular structure of the DGEBA and polyaminoamide used in this study.

In the second formulation, with 5 wt % ZP was used (epoxy-ZP coating). The epoxy resin was dissolved in acetone, to it was added the desired amount of zinc phosphate and the produced mixture was stirred at a speed of 1000 rpm for 20 min. The curing agent was then added slowly and the mixture of the three components was stirred at room temperature for 20 min. Produced formulations were applied by an air spray technique on 15CDV6 steel panels. Coated panels were then dried and cured at 60 °C for 1 h. The thickness of coatings was measured to be about 15–25 µm which was measured by a digital coating thickness gauge (Layercheck 750 USB FN, Erichsen Gmbh & Co. KG, Hemer, Germany). In order to get more accuracy for the overall coating thickness, it was measured triply for each sample and mean values are reported.

2.4. *Electrochemical Corrosion Tests and Surafce Characterization of Coating*

Electrochemical evaluation of coatings was carried out by electrochemical impedance spectroscopy (EIS) using a Potentiostat (SP-200, BioLogic, Knoxville, TN, USA) instrument. A three-electrode system was used for the electrochemical test used in this study as is shown in Figure 2. A saline solution with a 3 wt % NaCl concentration was used as a corrosive electrolyte. All electrochemical measurements were performed after 1 h immersion time.

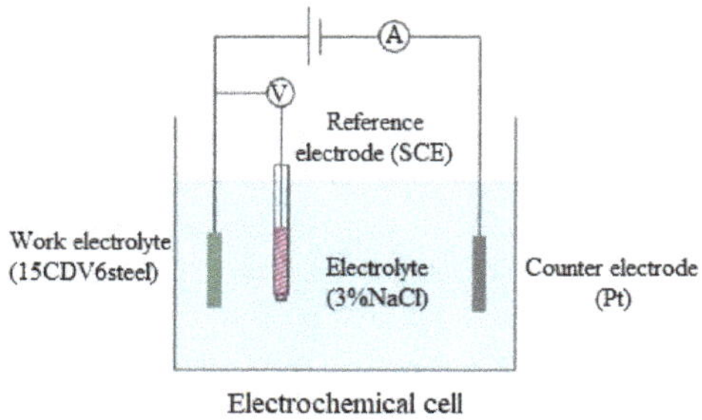

Figure 2. Schematic diagram of a three-electrode cell for electrochemical measurement.

In order to get more accuracy and reproducibility of experimental data, the electrochemical studies were performed triply at each tested sample and mean values are reported.

The EIS studies were carried out at a frequency range of 100 kHz to 10 mHz with an alternating current (AC) amplitude of ±10 mV at open circuit potential (OCP). The data points for each frequency were averaged to produce the 37 EIS data points.

Finally, the samples were exposed to the Neutral Salt Spray (NSS) test according to ASTM-B117 [22] before and after 4392 and 5865 h of exposure to the accelerated environment. The surface morphologies of the polymer coatings were characterized by scanning electron microscopy (SEM, S3000H, Hitachi, Tokyo, Japan), with an accelerating voltage of 20 kV.

3. Results and Discussion

3.1. Fourier-Transform Infrared Spectroscopy (FT-IR) Analysis

The Fourier Transform Infra-Red spectroscopy (FT-IR, FTS 6000, Digilab, Randolph, MA, USA) spectrum of polymeric epoxy cured with polyaminoamide on the 15CDV6 steel surface is shown in Figure 3.

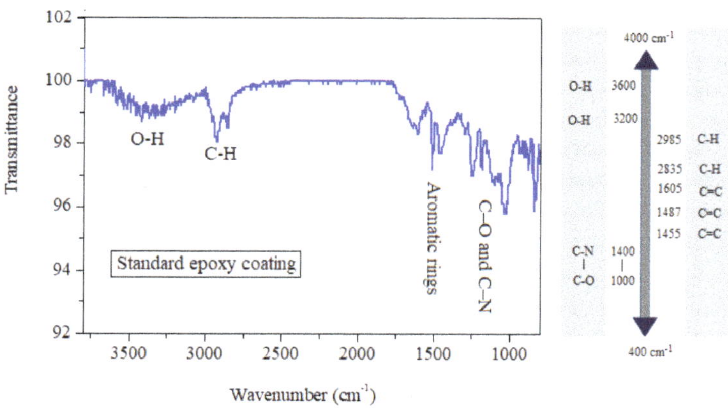

Figure 3. The Fourier-transform infrared spectroscopy (FT-IR) spectrum of polymeric epoxy cured with polyaminoamide on the 15CDV6 steel surface.

3.2. EIS Measurements

The Bode and Nyquist plots of standard epoxy coating and epoxy-ZP coating after 1464, 2928, and 4392 h exposure are shown in Figures 4 and 5.

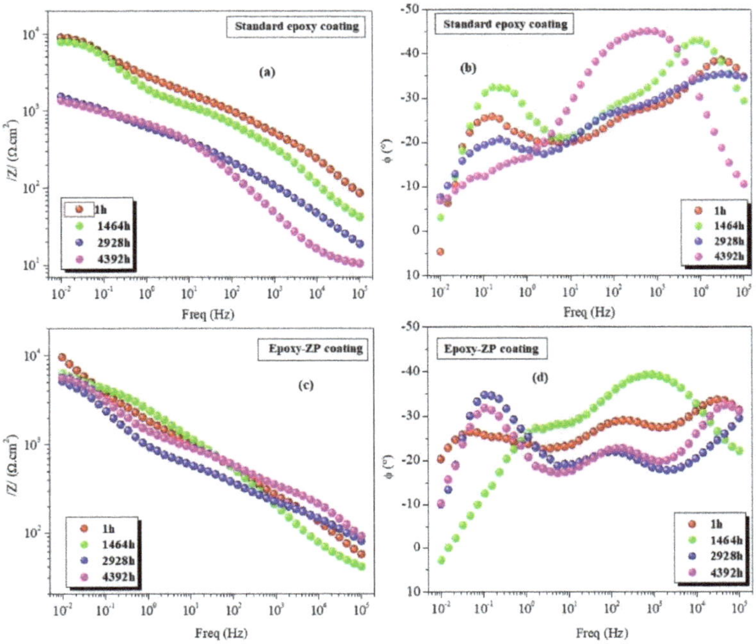

Figure 4. Bode magnitude plot (**a**,**c**) and Bode phase (**b**,**d**) plot obtained for the two polymer coatings applied on 15CDV6 steel with an area of 1 cm^2 before and after exposure for different times.

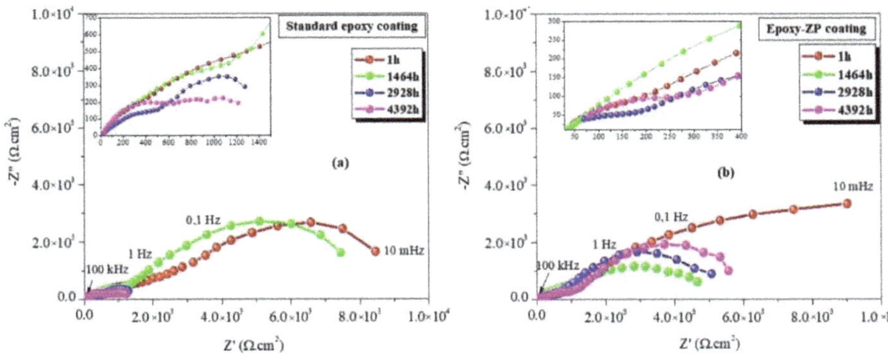

Figure 5. Nyquist plots obtained for the two polymer coatings applied on 15CDV6 steel before and after exposure for different times: (**a**) Standard epoxy coating and (**b**) Epoxy-ZP coating.

The Bode plot is a combination of the Bode magnitude plot and Bode phase plot which shows the magnitude of high frequency and low frequency regions due to coating capacitance and charge transfer process, respectively [23]. The impedance modulus (Bode magnitude) at low frequency ($|Z|_{0.01\ Hz}$) was obtained from EIS analysis and is presented in Figure 4. Results of Figure 4 clearly show the impedance diagrams of the standard epoxy coating and epoxy-ZP coating, they are characterized by two time constant.

The impedance modulus at 0.01 Hz was high, greater than 8 kΩ·cm^2. As can be seen from Figure 3, the total impedance of the epoxy coating in the high frequency region (9 × 10^3 kHz) after 1 h of immersion is slightly greater than that of the standard epoxy coating. The$|Z|_{0.01\ Hz}$ values of the standard epoxy coating started at 8.90 kΩ·cm^2 after 1 h immersion and dropped to 1.40 kΩ·cm^2 after

4392 h of exposure. The decrease in the impedance values with immersion time could be related to the dispersion of the corrosive electrolyte into the polymer coating, thus the barrier properties of the film are reduced. Further, the impedance for the epoxy-ZP coating is significantly higher than that for the standard epoxy coating, after 1 h of exposure it was 9.50 $k\Omega \cdot cm^2$ and then dropped to 5.90 $k\Omega \cdot cm^2$ after 4392 h of exposure. This shows the good barrier properties of epoxy-ZP coating, which could be attributed to the presence of the pigments as well as the epoxy resin. The pigments fill the micro holes and cuts in the epoxy coating, as a result of that the epoxy resin porosity diminishes [24–26].

The Nyquist plots of standard epoxy coating and epoxy-ZP coating after 1464, 2928, and 4392 h exposure are shown in Figure 5.

Bode phase plot ($\theta_{10\ kHz}$) at a high frequency is another parameter that was used to evaluate the protection performance of the polymer coatings. The $\theta_{10\ kHz}$ at a high frequency of standard epoxy coating and epoxy-ZP coating after a 1 h of immersion were about −31° and −36°, respectively. However, the $\theta_{10\ kHz}$ of standard epoxy coating decreases to about −30° after 4392 h of exposure. The drop in the phase angle could be due to the dispersion of the corrosive electrolyte into the polymer coating. Further, the $\theta_{10\ kHz}$ of the epoxy-ZP coating drops to about −33° after 4392 h of exposure. This shows the good barrier properties of epoxy-ZP coating.

The electrical equivalent circuit was used in this study as shown in Figure 6.

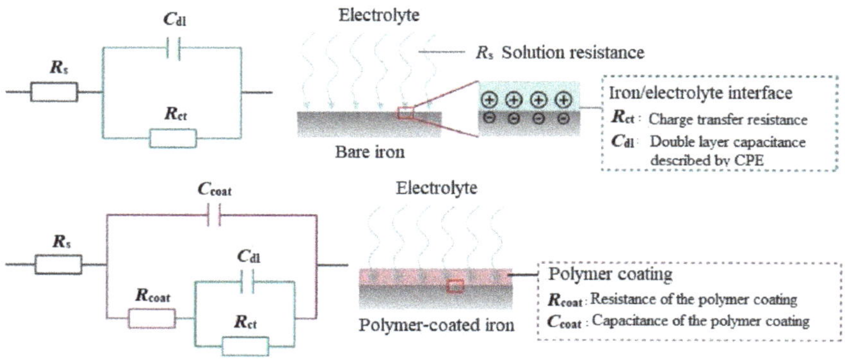

Figure 6. The equivalent electrical circuit used.

The Capacitance of the polymer coating values are calculated using Equation (1) [27]:

$$C_x = \frac{(R_x CPE)^{\frac{1}{n}}}{R_x} \quad (1)$$

where constant phase element (CPE) is non-ideal capacity and n is a capacity factor. The EIS data for the two polymer coatings are given in Table 3.

The results obtained are reported in Table 3, show that the resistance of the polymer coating (R_{coat}) and charge transfer resistance (R_{ct}) values for epoxy-ZP coating were greater than that for the standard epoxy coating, an indication of the superiority of the epoxy-ZP coating over the standard epoxy coating.

After 1 h of immersion, the R_{coat} and R_{ct} values of the epoxy-ZP coating were much higher than the one of the standard epoxy coating. After the first 1464 h of immersion the R_{coat} and R_{ct} values dropped for both the standard epoxy coating and the epoxy-ZP coating, a higher decrease was seen from the standard epoxy coating. After 2928 h of immersion, the R_{coat} and R_{ct} values of standard epoxy coating continued to decrease, while R_{coat} and R_{ct} values of epoxy-ZP coating actually showed an increase. As mentioned above, the observed results could be attributed to the penetration of the standard epoxy coating by water and electrolytes. While in the epoxy-ZP coating the penetration was blocked by the zinc phosphate micro-particles. The immersion period between 2928 and 4392 h showed a continuous decrease of in the R_{coat} and R_{ct} values of the standard epoxy coating, while the epoxy-ZP coating showed a continuous increase in the R_{coat} and R_{ct} values. Extending the immersion time to long period caused the water and corrosive electrolytes to reach the coating–metal interface causing an oxidation of active sites on the metal surface, leading to the formation of corrosion on the steel surface. While with the epoxy-ZP coating, the presence of ZP significantly improved the barrier properties of the polymer coating as mentioned earlier.

Table 3. The electrochemical parameters extracted from electrochemical impedance spectroscopy (EIS) data for two polymer coatings applied on 15CDV6 steel samples immersed in corrosive media for 1 h at room temperature.

Sample	Time (h)	Magnitude $\lvert Z\rvert_{0.01\ Hz}$ (kΩ·cm²)	Phase $-\theta_{10\ kHz}$ (deg)	R_s (Ω·cm²)	CPE$_{coat}$		R_{coat} (kΩ·cm²)	CPE$_{dl}$		R_{ct} (kΩ·cm²)	χ^2
					C_{coat} (μF/cm²)	n_{coat}		C_{dl} (μF/cm²)	n_{dl}		
Standard epoxy coating	1	8.90	31	25.01	65.6	0.36	6.90	15.0	0.74	6.62	0.21
	1464	7.64	33	11.54	30.3	0.51	2.13	20.0	0.73	6.60	0.18
	2928	1.60	33	03.39	35.1	0.36	0.73	10.4	0.66	1.07	0.03
	4392	1.40	30	09.37	48.8	0.68	0.03	51.2	0.19	2.00	0.07
Epoxy-ZP coating	1	9.50	36	16.7	24.0	0.30	6.16	89.2	1.00	16.60	0.11
	1464	6.65	43	24.6	53.3	0.52	3.42	10.4	0.66	4.21	0.11
	2928	5.10	36	6.11	33.9	0.26	2.25	19.3	0.77	7.53	0.09
	4392	5.90	33	68.9	29.7	0.22	2.50	10.0	1.00	8.50	0.17
	5136	7.60	38	45.8	10.1	0.62	2.93	38.6	0.33	13.6	0.52
	5496	6.90	33	51.0	9.55	0.64	2.89	51.4	0.38	6.46	0.82
	5856	2.30	27	12.7	22.0	0.51	1.93	20.5	0.48	7.94	0.67

The total impedance values extracted from the Nyquist and Bode plots were used to assess the stability of the coating during the exposure to the electrolytes for a long period of time [28]. It was also used to monitor the phenomena occurring on the coating–metal interface during the immersion of the coated steel in an electrolyte solution at various periods of time and for a long immersion time (4392 h).

The Bode (magnitude and phase) and Nyquist plots for the epoxy-ZP coating after exposure for different times are shown in Figure 7a–c.

The obtained electrochemical parameters are given in Table 3. The $\lvert Z\rvert_{0.01\ Hz}$ values of the epoxy-ZP coating starts at 7.60 kΩ·cm² after 5136 h of exposure and reaches 2.30 kΩ·cm² after 5856 h of exposure. On the other hand, the Bode phase plots $\theta_{10\ kHz}$ values of the epoxy-ZP coating begins at −38° after 5136 h immersion and reaches −27° after 5856 h of exposure. As shown in Table 3, the R_{coat} values of the epoxy-ZP coating showed a slight drop in the R_{coat} value from 2.93 to 1.93 kΩ·cm² after 5136 and 5856 h immersion, respectively. The slight drop in the R_{coat} values of the epoxy-ZP coating demonstrates that this coating has unique physical properties that allowed it to control the penetration of the electrolytes to the metal surface even after a long period of exposure to the corrosive electrolytes.

The corrosion rate is described by the charge transfer resistance (R_{ct}). For the epoxy-ZP coating it is observed that the value of R_{ct} was 13.6 kΩ·cm² after 5136 h of immersion in a corrosive electrolyte, while it is 6.46 and 6.01 kΩ·cm² after 5496 and 5856 h immersion, respectively.

In brief, zinc phosphate works by inhibiting the formation of a passivation layer containing Zn/P on the 15CDV6 steel, as schematically summarized in Figure 8.

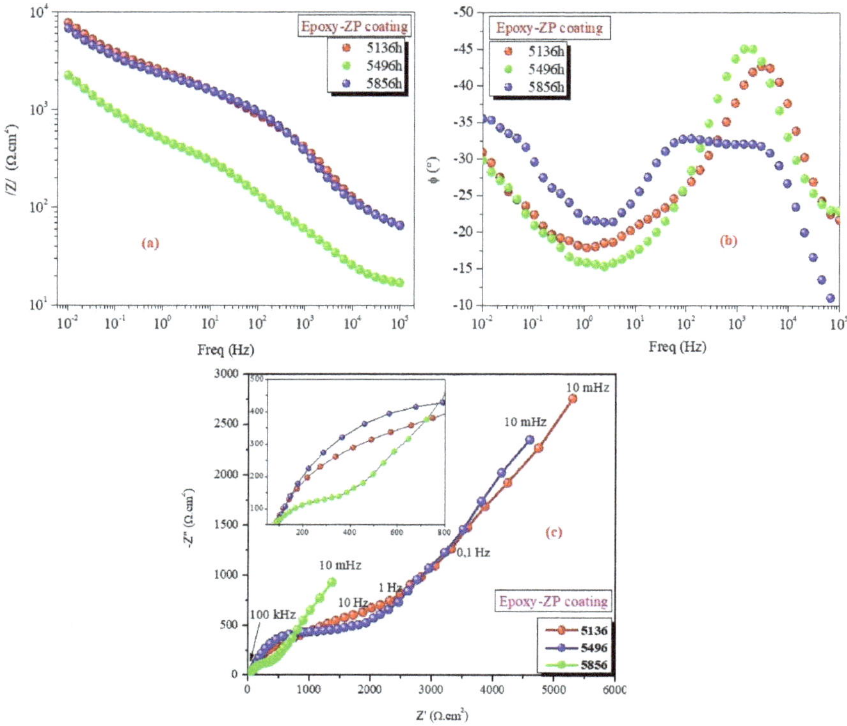

Figure 7. Bode (magnitude (**a**) and phase (**b**)) and Nyquist plots (**c**) obtained for the epoxy-ZP coating applied on 15CDV6 steel samples before and after exposure for different times.

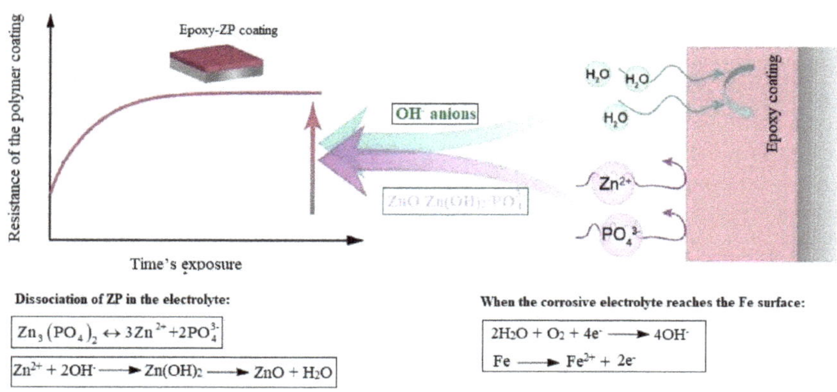

Figure 8. Schematic proposed for epoxy-ZP coating applied on 15CDV6 steel samples after exposure.

3.3. Surface Morphology of the Coatings

Scanning electron microscopy analysis was performed to investigate the surface morphology of the two evaluated polymer coatings: Standard epoxy coating and the epoxy-ZP coating. The coating surface exposed to an electrolyte solution in a spray test chamber was also evaluated at various times of exposure. Results are shown in Figures 9 and 10.

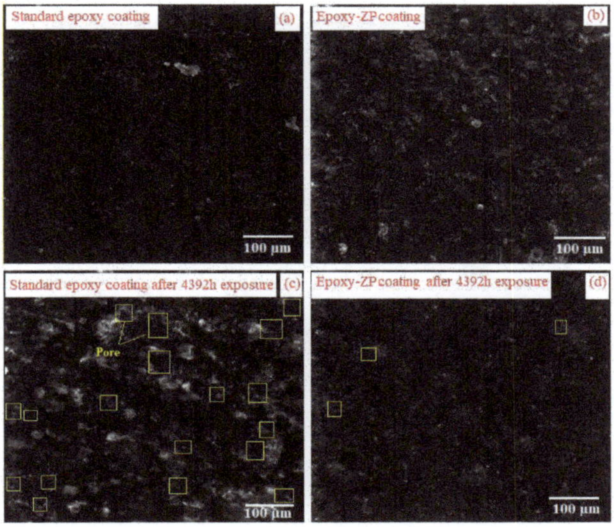

Figure 9. SEM-images of the standard epoxy coating (**a**) and epoxy-ZP coating (**b**) applied on 15CDV6 steel before and after 4392 h exposure (**c**,**d**, respectively).

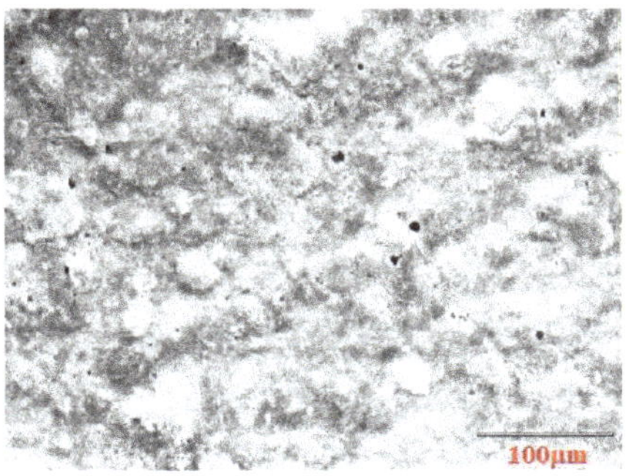

Figure 10. SEM image for 15CDV6 steel coated with epoxy-ZP coating after 5856 h of exposure.

As shown in Figure 9a, before the exposure, the standard epoxy coating surface is uniform, smooth and free from defects with a low number of aggregates. The polymer coatings containing 5 wt % zinc phosphate (Figure 9b), shows that zinc phosphate is well dispersed in the coating layer, and is more uniform and homogeneous than the standard coating.

After 4392 h of exposure, the standard epoxy coating (Figure 9c) showed porous morphology and the corrosion site can be observed on the 15CDV6 steel surface. However, the surface of the epoxy-ZP coating showed fewer scratches and corrosion sites Figure 9d. As mentioned above, the results confirm that the addition of zinc phosphate facilitated the formation of a passive phosphate barrier film on the 15CDV6 steel substrate and inhibited the penetration of corrosive electrolyte ions (H_2O, O_2, and Cl^-) to the 15CDV6 steel. It also enhances the anticorrosion process after 4392 h exposure.

The surface morphology of the epoxy-ZP coating was also evaluated after 5856 h of exposure. The obtained SEM micrograph is presented in Figure 10. From Figure 10 it can be obviously seen that the surface morphology of epoxy coating containing ZP is significantly changed. The surface became more porous and the epoxy-ZP coating surface can be seen in the figure. These observations show that surface corrosion sites also developed with the epoxy coating containing ZP.

4. Conclusions

In this study, two polymer-based coatings (standard epoxy coating) and (epoxy-ZP coating) were evaluated as an anticorrosive for steel 15CDV6. The studies were performed on coated samples before and after exposure. The experimental results represented by EIS results indicate that the epoxy-ZP coating exhibited excellent physical barrier properties and high pore resistance value. So, the dual component coating of the present work is a nontoxic, ecofriendly coating with a high potential in marine corrosion protection. A surface study of the polymer coating by SEM showed a homogeneous distribution of zinc phosphate in the polymer coating.

Author Contributions: Methodology, O.D. and S.J.; Results Discussion, B.K. and G.H.; Validation and Supervision, A.E.H.; Writing—Review and Editing, S.J. and O.D; Editing and Financial Support, A.D.

Funding: This research received no external funding.

Acknowledgments: The authors would like to thank An-Najah National University and Ibn Tofail University for providing the necessary equipment and chemicals for getting this work accomplished. The authors would also like to thank the laboratory of metallurgical analysis, Cetim Maroc Dévelopement and quality control laboratory, Casablanca Aeronautics Group Figeac Aero. Aeronautical Technopole of Nouaceur, Mohammed V-Casablanca Airport, Morocco. Finally, we would like to thank the Palestine Ministry of Higher Education and Research Scientific

Conflicts of Interest: The authors declare no conflict of interest.

References

1. Dagdag, O.; El Harfi, A.; El Gouri, M.; Safi, Z.; Jalgham, R.T.; Wazzan, N.; Verma, C.; Ebenso, E.E.; Kumar, U.P. Anticorrosive properties of Hexa (3-methoxy propan-1, 2-diol) cyclotri-phosphazene compound for carbon steel in 3% NaCl medium: Gravimetric, electrochemical, DFT and Monte Carlo simulation studies. *Heliyon* **2019**, *5*, e01340. [CrossRef] [PubMed]
2. Dagdag, O.; El Gana, L.; Hamed, O.; Jodeh, S.; El Harfi, A. Anticorrosive formulation based of the epoxy resin–polyaminoamide containing zinc phosphate inhibitive pigment applied on sulfo-tartaric anodized AA 7075-T6 in NaCl medium. *J. Bio-and Tribo-Corros.* **2019**, *5*, 25. [CrossRef]
3. Indumathi, S.N.; Vasudevan, T.; Sundarrajan, S.; Rao, B.S.; Murthy, C.V.S.; Yadav, D.R. Cadmium-and chromate-free coating schemes for corrosion protection of 15CDV6 steel. *Met. Finish.* **2011**, *109*, 15–21. [CrossRef]
4. Ramkumar, P.; Prakash, F.G.; Karthikeyan, M.K.; Gupta, R.K.; Muthupandi, V. Development of copper coating technology on high strength low alloy steel filler wire for aerospace applications. *Mater. Today Proc.* **2018**, *5*, 7296–7302. [CrossRef]
5. Dagdag, O.; Essamri, A.; El Gana, L.; El Bouchti, M.; Hamed, O.; Cherkaoui, O.; Jodeh, S.; El Harfi, A. Synthesis, characterization and rheological properties of epoxy monomers derived from bifunctional aromatic amines. *Polym. Bull.* **2018**, 1–15. [CrossRef]
6. Dagdag, O.; Hamed, O.; Erramli, H.; El Harfi, A. Anticorrosive performance approach combining an epoxy polyaminoamide–zinc phosphate coatings applied on sulfo-tartaric anodized aluminum alloy 5086. *J. Bio-and Tribo-Corros.* **2018**, *4*, 52. [CrossRef]
7. Dagdag, O.; Hsissou, R.; Berisha, A.; Erramli, H.; Hamed, O.; Jodeh, S.; El Harfi, A. Polymeric-based epoxy cured with a polyaminoamide as an anticorrosive coating for aluminum 2024-T3 surface: experimental studies supported by computational modeling. *J. Bio-and Tribo-Corros.* **2019**, *5*, 58. [CrossRef]
8. Pourhashem, S.; Vaezi, M.R.; Rashidi, A.; Bagherzadeh, M.R. Exploring corrosion protection properties of solvent based epoxy-graphene oxide nanocomposite coatings on mild steel. *Corros. Sci.* **2017**, *115*, 78–92. [CrossRef]
9. Dagdag, O.; El Harfi, A.; Essamri, A.; El Bachiri, A.; Hajjaji, N.; Erramli, H.; Hamed, O.; Jodeh, S. Anticorrosive performance of new epoxy-amine coatings based on zinc phosphate tetrahydrate as a nontoxic pigment for carbon steel in NaCl medium. *Arab. J. Sci. Eng.* **2018**, *43*, 5977–5987. [CrossRef]
10. Dagdag, O.; El Harfi, A.; Essamri, A.; El Gouri, M.; Chraibi, S.; Assouag, M.; Benzidia, B.; Hamed, O.; Lgaz, H.; Jodeh, S. Phosphorous-based epoxy resin composition as an effective anticorrosive coating for steel. *Int. J. Ind. Chem.* **2018**, *9*, 231–240. [CrossRef]

11. Zhang, Z.; Zhang, W.; Li, D.; Sun, Y.; Wang, Z.; Hou, C.; Chen, L.; Cao, Y.; Liu, Y. Mechanical and anticorrosive properties of graphene/epoxy resin composites coating prepared by in-situ method. *Int. J. Mol. Sci.* **2015**, *16*, 2239–2251. [CrossRef] [PubMed]
12. Jirasutsakul, I.; Paosawatyanyong, B.; Bhanthumnavin, W. Aromatic phosphorodiamidate curing agent for epoxy resin coating with flame-retarding properties. *Prog. Org. Coat.* **2013**, *76*, 1738–1746. [CrossRef]
13. Parhizkar, N.; Ramezanzadeh, B.; Shahrabi, T. Corrosion protection and adhesion properties of the epoxy coating applied on the steel substrate pre-treated by a sol-gel based silane coating filled with amino and isocyanate silane functionalized graphene oxide nanosheets. *Appl. Surf. Sci.* **2018**, *439*, 45–59. [CrossRef]
14. Sababi, M.; Terryn, H.; Mol, J.M.C. The influence of a Zr-based conversion treatment on interfacial bonding strength and stability of epoxy coated carbon steel. *Prog. Org. Coat.* **2017**, *105*, 29–36. [CrossRef]
15. Liu, X.; Xiong, J.; Lv, Y.; Zuo, Y. Study on corrosion electrochemical behavior of several different coating systems by EIS. *Prog. Org. Coat.* **2009**, *64*, 497–503. [CrossRef]
16. Naderi, R.; Attar, M.M. The role of zinc aluminum phosphate anticorrosive pigment in protective performance and cathodic disbondment of epoxy coating. *Corros. Sci.* **2010**, *52*, 1291–1296. [CrossRef]
17. Mousavifard, S.M.; Nouri, P.M.; Attar, M.M.; Ramezanzadeh, B. The effects of zinc aluminum phosphate (ZPA) and zinc aluminum polyphosphate (ZAPP) mixtures on corrosion inhibition performance of epoxy/polyamide coating. *J. Ind. Eng. Chem.* **2013**, *19*, 1031–1039. [CrossRef]
18. Blustein, G.; Deyá, M.C.; Romagnoli, R.; Del Amo, B. Three generations of inorganic phosphates in solvent and water-borne paints: A synergism case. *Appl. Surf. Sci.* **2005**, *252*, 1386–1397. [CrossRef]
19. Deyab, M.A.; Ouarsal, R.; Al-Sabagh, A.M.; Lachkar, M.; El Bali, B. Enhancement of corrosion protection performance of epoxy coating by introducing new hydrogenphosphate compound. *Prog. Org. Coat.* **2017**, *107*, 37–42. [CrossRef]
20. Cotting, F.; Aoki, I.V. Smart protection provided by epoxy clear coating doped with polystyrene microcapsules containing silanol and Ce (III) ions as corrosion inhibitors. *Surf. Coat. Technol.* **2016**, *303*, 310–318. [CrossRef]
21. Deyab, M.A. Effect of carbon nano-tubes on the corrosion resistance of alkyd coating immersed in sodium chloride solution. *Prog. Org. Coat.* **2015**, *85*, 146–150. [CrossRef]
22. *ASTM B117-18 Standard Practice for Operating Salt Spray (Fog) Apparatus*; ASTM: West Conshohocken, PA, USA, 2019.
23. Deflorian, F.; Rossi, S.; Fedel, M.; Motte, C. Electrochemical investigation of high-performance silane sol–gel films containing clay nanoparticles. *Prog. Org. Coat.* **2010**, *69*, 158–166. [CrossRef]
24. Montemor, M.F.; Snihirova, D.V.; Taryba, M.G.; Lamaka, S.V.; Kartsonakis, I.A.; Balaskas, A.C.; Kordas, G.C.; Tedim, J.; Kuznetsova, A.; Zheludkevich, M.L.; et al. Evaluation of self-healing ability in protective coatings modified with combinations of layered double hydroxides and cerium molibdate nanocontainers filled with corrosion inhibitors. *Electrochim. Acta* **2012**, *60*, 31–40. [CrossRef]
25. Ammar, S.; Ramesh, K.; Vengadaesvaran, B.; Ramesh, S.; Arof, A.K. Amelioration of anticorrosion and hydrophobic properties of epoxy/PDMS composite coatings containing nano ZnO particles. *Prog. Org. Coat.* **2016**, *92*, 54–65. [CrossRef]
26. Yu, D.; Wen, S.; Yang, J.; Wang, J.; Chen, Y.; Luo, J.; Wu, Y. RGO modified ZnAl-LDH as epoxy nanostructure filler: A novel synthetic approach to anticorrosive waterborne coating. *Surf. Coat. Technol.* **2017**, *326*, 207–215. [CrossRef]
27. Liu, X.; Gu, C.; Wen, Z.; Hou, B. Improvement of active corrosion protection of carbon steel by water-based epoxy coating with smart CeO_2 nanocontainers. *Prog. Org. Coat.* **2018**, *115*, 195–204. [CrossRef]
28. Roselli, S.N.; Romagnoli, R.; Deyá, C. The anti-corrosion performance of water-borne paints in long term tests. *Prog. Org. Coat.* **2017**, *109*, 172–178. [CrossRef]

coatings

MDPI

Article

Inclusion of 5-Mercapto-1-Phenyl-Tetrazole into β-Cyclodextrin for Entrapment in Silane Coatings: An Improvement in Bronze Corrosion Protection

Cecilia Monticelli [1,*], Giancarlo Fantin [2], Graziano Di Carmine [2], Federica Zanotto [1] and Andrea Balbo [1]

[1] Centro di Studi sulla Corrosione e Metallurgia "A. Daccò", Dipartimento di Ingegneria, Università di Ferrara, 44121 Ferrara, Italy
[2] Dipartimento di Scienze Chimiche e Farmaceutiche, Università di Ferrara, 44121 Ferrara, Italy
* Correspondence: mtc@unife.it; Tel.: +39-0532-455136

Received: 23 July 2019; Accepted: 8 August 2019; Published: 10 August 2019

Abstract: The corrosion protection of coatings can be reinforced by the addition of entrapped corrosion inhibitors. β-cyclodextrin (β-CD) can form inclusion complexes with small inhibiting organic molecules that, when entrapped in coatings, allow the inhibitor release and adsorption at corrosion initiation sites. In this paper, several Nuclear Magnetic Resonance (NMR)-based experiments (e.g., Complexation-Induced Shifts (CIS), NMR titration, Diffusion-Ordered Spectroscopy (DOSY)) were performed to study the stability and geometry of a complex formed by β-cyclodextrin with 5-mercapto-1-phenyl-tetrazole (MPT). The complex was also detected by Electrospray Ionization (ESI) mass spectrometry and characterized by Fourier Transform Infrared (FTIR) spectra. Its influence on the protectiveness of a silane coating against bronze corrosion was evaluated in plain (AR) and concentrated (ARX10) synthetic acid rain, under different exposure conditions. In particular, the time evolution of the polarization resistance values during 20 days in ARX10 and the polarization curves recorded at the end of the immersions evidenced a higher protectiveness of the coating with the β-CD–MPT complex in comparison to that containing only MPT or only β-CD. The cyclic AR spray test carried out on coated bronze coupons with cross-cut scratches evidenced the absence of underfilm corrosion starting from the scratches only in the complex-containing coating.

Keywords: corrosion protection; smart coatings; inclusion complex; complex stability; bronze

1. Introduction

The corrosion protection of an organic coating depends largely on the intrinsic barrier properties of the polymeric film towards oxygen, water, and aggressive species, but can be reinforced by entrapped corrosion inhibitors [1–3]. In the case of bronze artworks exposed outdoors, commercial coatings such as Incralac® and Soter® contain benzotriazole (BTA) that operates in the dual functions of an inhibitor of bronze corrosion and an anti-UV additive [4]. In these coatings, the inhibitor dissolves in the electrolyte (in the rain) and penetrates through the coating, so producing an inhibited solution that exerts a protective action at the coating/metal interface. In general, the positive influence of direct inhibitor addition in coatings may be limited by solubility problems, by a decrease in barrier properties and adherence to the substrate, and by the rapid leaching of small inhibitor molecules induced by rainfalls, so determining a fast drop of the overall protection performance [5].

In recent years, great efforts were devoted to overcoming these problems and increasing the coating durability by encapsulating corrosion inhibitors in coatings through the adoption of suitable carriers, which make them more compatible with the coating network [6,7]. Different carrier types were investigated, such as inhibitor-filled porous particles, or nanocapsules, coated by polyelectrolytes [8–10] or layered anion-exchange particles containing intercalated corrosion inhibiting anionic species [11]. The former polyelectrolyte-coated carriers can release inhibitor molecules due to local variations of solution pH linked to the onset of corrosion phenomena, while the latter carriers with intercalated inhibiting anionic species limit the access of aggressive anions, such as chlorides, to the metal substrate and release corrosion inhibiting species by ion-exchange phenomena.

Beside these carriers, cyclodextrins (CD) were also used to host corrosion inhibitors due to their complexing capability towards small organic molecules with corrosion inhibition properties. Cyclodextrins (CDs) [12] are water-soluble macrocyclic oligosaccharides consisting of at least six α -D-glucopyranose units linked via α (1–4) glycosidic bonds. The most common ones contain 6, 7 and 8 glucopyranose residues and are known as α-, β-, and γ-CD, respectively. CDs are among the most widely used host molecules thanks to their unique and specific structure that creates an internal cavity that is less hydrophilic than the external aqueous environment and therefore able to accommodate a large variety of hydrophobic molecules inside it [13]. The strong propensity of CDs to form inclusion complexes has been exploited not only for the production of smart coatings for corrosion protection, but also in many other fields of science, for example, as drug transport systems, [14], to increase the solubility of some chemical species [15], in separation technology [16], and in other areas [17].

Addition of complexes of α- and β-CDs with dibenzylthiourea (DBT) in acid solutions improved the corrosion resistance of carbon steel due to the enhanced solubility of the complexes in comparison to that of DBT alone [18]. The controlled release of BTA from β-CD–BTA complexes was investigated in order to achieve effective bronze corrosion protection in chloride solutions. The use of the complex instead of pure BTA was intended to reduce the toxicity of the additives used for corrosion protection [19]. Moreover, complexes of β-CD and γ-CD with mercaptobenzothiazole (MBT), mercaptobenzimidazole (MBI), mercaptobenzimidazole sulfonate, and thiosalicylic acid [20–23] were incorporated in coatings and improved the corrosion resistance of Al alloys and zinc. In fact, these complexes represent a reservoir of corrosion inhibitor molecules that, at the onset of corrosion phenomena and in correspondence of regions of coating delamination, tend to adsorb on the metal surfaces, shifting the complexation equilibrium towards the release of further inhibitor molecules.

In this research, the effective formation and the stability of a complex between β-CD and 5-mercapto-1-phenyl-tetrazole (MPT) (Figure 1) were assessed. Among non-toxic corrosion inhibitors [24–26], MPT was chosen because of its outstanding inhibiting properties towards copper and bronze corrosion [27–29]. Its complex with β-CD was incorporated in a 3-mercapto-propyl-trimethoxy-silane (PropS-SH) coating, and the protectiveness of the obtained coating was assessed on bronze during both immersions in concentrated acid rain and exposures to alternated acid rain spray. These tests were also performed on plain silane coating and coatings with only β-CD or only MPT additions. For the continuous immersion tests, thin sub-micrometric silane films prepared by the dip coating method were used, in order to better differentiate their protectiveness. During the alternated acid rain spray test, the self-healing properties were assessed on thicker coatings (about 5 μm) produced by spraying after introduction of cross cut scratches. Spraying and brushing are the application methods most commonly adopted by restorers [30] and produce coatings more representative of those actually applied in the field.

It is important to stress that according to occupational hazard tests, the silane formulations here investigated were less toxic to restorers than Incralac® (as both pure product and ready to use 30% solution in nitro diluent) [31].

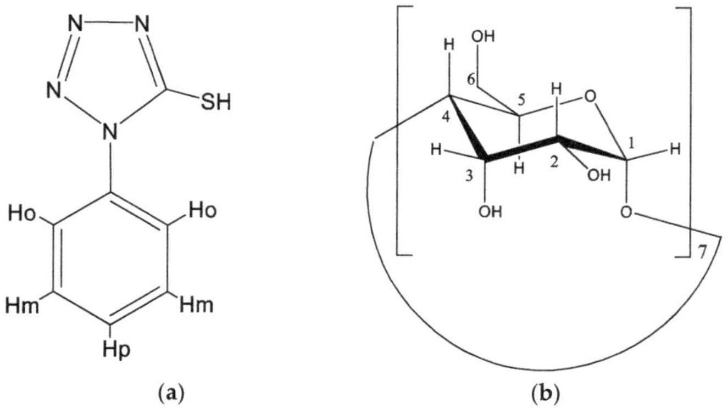

Figure 1. Molecular structure of (**a**) 5-mercapto-1-phenyl-tetrazole (MPT) and (**b**) β-cyclodextrin (β–CD). β–CD is a CD type constituted by 7 glucopyranoside units linked by 1,4 glycosydic bonds.

2. Materials and Methods

2.1. Chemicals, Aggressive Environment and Alloy

All reagents and solvents used in this study were purchased from commercial sources. In particular, the chemicals used for the inhibitor complex and the coating production were β-cyclodextrin (β-CD, ≥ 97% purity), 5-mercapto-1-phenyl-tetrazole (MPT, 98% purity), and 3-mercapto-propyl-trimethoxy-silane (PropS-SH, purity 95%), all purchased from Sigma-Aldrich (Darmstadt, Germany).

The coating protectiveness was tested on as-cast bronze with composition 91.9 Cu, 2.4 Sn, 1.0 Pb, Zn 2.9, 0.8 Sb, wt.%, and a microstructure reproducing those of Renaissance bronze artefacts with cored dendrites of Cu-solid solution characterized by Sn and Sb local enrichment and also including Pb globules in the interdendritic spaces, as reported in previous papers [29,31–33].

Concerning the environments where the coating protectiveness was assessed, the cyclic acid rain (AR) spray test was performed using a synthetic AR, prepared with Sigma-Aldrich ACS reagents, according to the recipe reported in [34] and containing the following ion concentrations: SO_4^{2-} 1.90 mg· L^{-1}, Cl^- 1.27 mg· L^{-1}, NO_3^- 4.62 mg· L^{-1}, CH_3COO^- 0.23 mg· L^{-1}, $HCOO^-$ 0.05 mg· L^{-1}, NH_4^+ 1.05 mg· L^{-1}, Ca^{2+} 0.34 mg· L^{-1}, Na^+ 0.53 mg· L^{-1}, and pH 4.35. During the electrochemical tests, accelerated corrosion conditions were obtained by using tenfold concentrated AR (ARX10, pH = 3.3) at 30 °C.

2.2. [β-CD–MPT] Complex Stability Analysis

2.2.1. Nuclear Magnetic Resonance (NMR) Measurements

The NMR spectra were recorded in D_2O solution using 5 mm tubes, at 296 K, with a Varian Mercury Plus 400 (Palo Alto, CA, USA), operating at 400 (^1H) and 100 MHz (^{13}C), respectively. The chemical shifts were referenced to the DOH signal: δ (H) 4.65 ppm. The 1D-Rotating frame Overhauser effect spectroscopy (ROESY) NMR spectra were acquired using standard pulse sequences from the Varian library. The relaxation delay between successive pulse cycles was 1.0 s.

The Diffusion-Ordered Spectroscopy (DOSY) experiments were performed using the Dosy Bipolar Pulsed Pair STimulated Echo (DBPPSTE) pulse sequence [35] from the Varian library, using 15 different gradient values varying from 2% to 95% of the maximum gradient strength. A 500 ms diffusion time was chosen, and the gradient length was set to 2.0 ms. The analysis of all NMR spectra was performed with MestreNova (by Mestrelab Research, S.L., Santiago de Compostela, Spain), version: 6.0.2–5475 and for the DOSY analysis, the Baysian DOSY transform from MestreNova, version: 6.0.2–5475 was used.

2.2.2. H NMR Titration

The following two solutions were prepared in D_2O: Solution A: 2.8 mM MPT. Solution B: 2.8 mM MPT and 12.0 mM β-CD. A 0.8 mL aliquot of solution A was placed in a 5 mm NMR tube. A measured amount of solution B was added, changing the molar fraction of the host to about 0, 0.39, 0.71, 0.98, 1.21, 1.68, 2.02 and 2.29. Spectra were recorded after each addition. The chemical shift variation of the guest signals was collected, and the binding constants β (as log K) were calculated by the curve fitting method [36] using the commercial HypNMR2008 [37] program (details are given in the Supplementary Information (SI) file).

2.2.3. Electrospray Ionization (ESI) Mass Spectra

ESI mass spectra were obtained using an LCQ Duo (ThermoQuest, San Jose, CA, USA) in negative-ion mode. Instrumental parameters: capillary voltage −10 V, spray voltage 4.50 kV, mass scan range was from m/z 100 to 2000 amu, for 30,000 ms scan time; N^2 was used as the sheath gas. The samples were injected into the spectrometer through a syringe pump at a constant flow rate of 8 mL/min.

2.2.4. Fourier Transform Infra-Red (FTIR) Analysis

Diffuse reflectance FTIR spectra were recorded on β-CD–MPT complex powder and on MPT and β-CD powders, as references. The instrument used was a Thermo-Scientific Nicolet iS50 spectrometer (Thermo Fisher Scientific, Waltham, MA, USA), operating in dry CO_2-free air flow generated by a Balston 75-52 unit. It was equipped by a deuterated triglycine sulfate (DTGS) detector, which allowed for the investigation of the 4000–400 cm^{-1} wavenumber region with a resolution of 4 cm^{-1}.

2.3. *Silane Coating Production*

Silane hydrolysis was carried out by dissolving PropS-SH in a hydroalcoholic solution (90/5/5 v/v ethanol/water/PropS-SH), acidified to pH 4 by the addition of some drops of diluted sulphuric acid solution, according to the methodology refined in previous research works [29,38–40].

Plain PropS-SH coatings were directly produced from this solution after 24 h room-temperature ageing. The coating additives were introduced in the aged silane hydroalcoholic solution; in particular, 3 mL aqueous solutions of either 5.94 mM MPT or 5.94 mM β-CD or 5.94 mM β-CD + 5.94 mM MPT were added to 30 mL of silane solution, so that the final molar concentration of the additives in the coating formulations was 0.54 mM. These solutions were sonicated for 3 min and then applied to the substrate either by dip coating (1 h immersion and then fast withdrawal, reaching a final coating thickness of about 300 nm; for accelerated electrochemical tests) or by spraying (to reach a final constant coating specific weight of 6 ± 1 $g \cdot m^{-2}$ and thickness of 5 ± 1 μm; for cyclic AR spray tests). Finally, the coatings were cured for 24 h at 50 °C. This low temperature curing was compatible with the requirements for cultural heritage bronze artworks.

2.4. *Silane Coating Protectiveness*

Electrochemical tests were performed under accelerated corrosion conditions, that is in ARX10 (pH 3.3) at 30 °C, on thin dip-coated bronze electrodes. As a reference, tests were also carried out on bare bronze electrodes.

The evolution of corrosion conditions was monitored over 20 days of immersion by Electrochemical Impedance Spectroscopy (EIS, performed by a PARTSTAT 2273, from Ametek, Berwyn, PA, USA) tests performed at intervals, under the following experimental conditions: corrosion potential (E_{cor}) ± 10 mV rms, 10 kHz–1 mHz frequency range and 10 frequencies/decade. Polarization resistance (R_p) values were estimated from the spectra in the Nyquist form, as the difference between the limit of the real part of the impedance at frequency tending to 0 (R_p') and the solution resistance (R_s) value ($R_p = R_p' - R_s$) [40]. R_p values are inversely proportional to the corrosion currents (i_{cor}), as indicated by the Stern and Geary relationship [41]: $R_p = \frac{B}{i_{cor}}$, with B a constant depending on the Tafel slopes of the anodic and cathodic polarization curves. The time evolutions of average R_p and E_{cor} values were obtained from triplicate experiments.

Ohmic drop-compensated polarization curves were collected at the end of the 20 day immersion period. Separate anodic and cathodic potential scans, always starting from E_{cor}, were carried out at a rate of 0.1667 $mV \cdot s^{-1}$. These tests were performed in triplicate and representative curves were reported.

The self-healing capability of PropS-SH coatings was assessed by exposing coated coupons with cross cut scratches to a cyclic AR spray test at 35 °C for 4 weeks. Each cycle consisted in 8 h spraying and 16 h waiting. During the test, the coupons were supported with an angle of 30° from the vertical. Micrographs documented the extent of the final corrosion attack.

3. Results

3.1. NMR Studies

To verify the effective formation of a host–guest complex between β-CD and MPT we mainly employed the NMR spectroscopy, already widely used to study inclusion complexes of CD [42–44].

The ^1H NMR spectrum in D_2O of a sample containing an equimolar amount of β-CD and MPT showed induced chemical shifts (Δδ) for all the protons of the host as well as for most of the guest protons (Figure 2). The complexation-induced shifts (CIS) observed for the internal protons of β-CD (H3, H5) were remarkably larger than those displayed by the external ones (H-1, H-2, and H-4) (Table 1), thus indicating that the guest likely interacts with the inner cavity of β-CD.

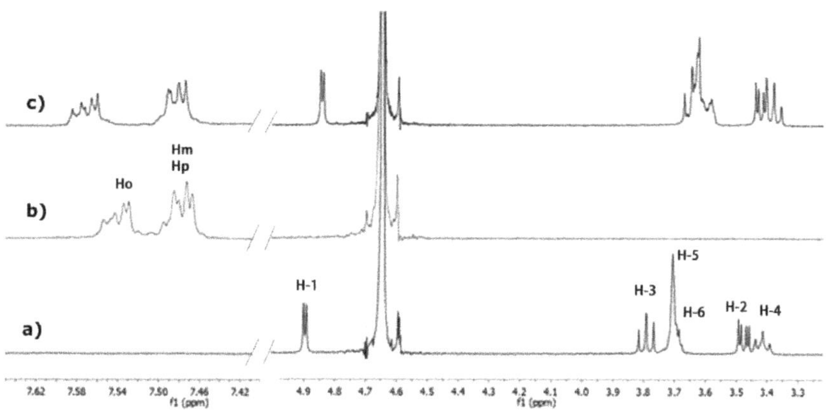

Figure 2. The ^1H NMR spectra in D_2O for solutions of (**a**) β-CD; (**b**) MPT; (**c**) equimolar amount of β-CD and MPT.

Table 1. The ^1H-NMR chemical shifts (δ, ppm) for H protons of β-CD alone, MPT alone, and complexation induced shifts (CIS = $\delta_{complex} - \delta_{free}$) of equimolar amounts of them in D_2O at 23 °C.

β-CD	Protons	H-1	H-2	H-3	H-4	H-5	H-6
δ alone		4.90	3.48	3.80	3.42	3.71	3.69
CIS		−0.06	−0.06	−0.15	−0.04	−0.09	−1.1
MPT	**Protons**	**H-o** [a]	**H-m** [a,b]	**H-p** [a,b]			
δ alone		7.54	7.48	7.48			
CIS		0.04	0.01	0.01			

[a] The center of the multiplet was taken into account; [b] overlapping signals.

The small CIS observed for the ortho-protons of the phenyl moiety of MPT (0.04 ppm, Table 1) was also particularly significant, indicating that a part of the phenyl ring of MPT entered into the cavity of β-CD. It is worthy of note that no new peaks appeared in the spectrum, signifying that the inclusion of MPT in β-CD is a fast exchange process that takes place on the NMR timescale.

In order to gather information on the sites of binding we carried out a series of monodimensional ROESY-1D experiments [45] that provided only a small nuclear Overhauser effect (NOE) on the inner H3 proton of β-CD when ortho-H phenyl protons of MPT were irradiated.

The experimental observations collected up to this point were compatible with the inclusion structure of Figure 3, which was in rapid equilibrium with the two separate molecules. In fact, the large CIS exhibited by inner protons H3 and H5 of β-CD upon the addition of MPT clearly indicated a deep insertion of the host into the hydrophobic cavity of the β-CD. However, the lack of dipolar interactions between the ortho- and meta-protons of the aromatic ring of MPT and the H5 proton of β-CD excluded the complete insertion of the phenyl moiety into the β-CD cavity. On the other hand, the weak rotating-frame Overhauser effect (ROE) existing between the ortho-protons of MPT and the H3 proton of β-CD suggested that a partial insertion of the phenyl moiety of MPT occurred at the wide rim of β-CD. Finally, the upfield shift of the H-6 protons of the β-CD (see Table 1) could be justified by the partial protrusion of the guest (tetrazole moiety) from the narrow rim of the β-CD (Figure 3) [46].

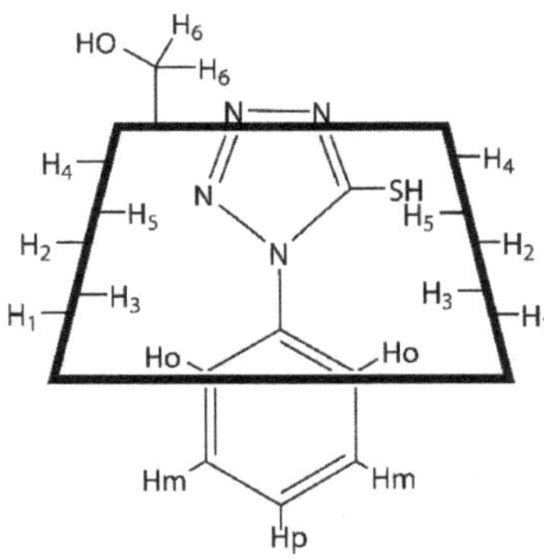

Figure 3. Proposed geometry for the inclusion of MPT into β-CD as deduced from the CIS and ROESY-1D experiments.

To investigate the strength of complexation we carried out a ^{1}H NMR titration of MPT with β-CD [36]. The procedure adopted in the titration allowed us to operate with a constant guest concentration ([MPT] = constant) during the whole experiment.

The binding isotherm relative to the ortho-protons of the phenyl moiety of MPT is depicted in Figure 4. The value of the stability constant β as log K_a of the β-CD–MPT inclusion complex was calculated by the curve fitting method [36], using the commercial HypNMR2008 [37] program (details are given in SI) and was found to be equal to 2.93 M^{-1} (K_a = 851 M^{-1}).

The 1:1 stoichiometry of the complexation adequately described the binding data obtained from the NMR titration and, on the other hand, the physically unrealistic binding parameters (some negative K_s) when the 1:2 or 2:1 models were applied to the NMR titration. This confirmed that the 1:1 stoichiometry was dominating in the investigated concentration range. The 1:1 stoichiometry of the complex was further confirmed by ESI-Mass Spectrometry (ESI-MS).

DOSY spectroscopy [37,42–47] experiments were also carried out to confirm, qualitatively and quantitatively, the intermolecular interactions between MPT and β-CD in solution. The DOSY technique allowed for the determination of the individual self-diffusion coefficients (D) in multicomponent systems that directly reflected the association behavior of the interacting species [48].

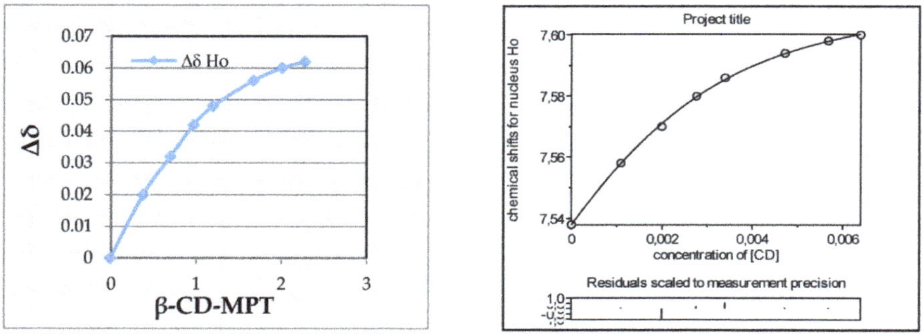

(**a**) (**b**)

Figure 4. The ^{1}H NMR titration of MPT with β-CD: (**a**) Chemical shift change of the ortho-protons of the phenyl moiety of MPT with increasing β-CD concentration. Positive values mean downfield shifts. The small complexation induced shift (CIS) observed for the ortho-protons of the phenyl moiety of MPT (0.04 ppm, see Table 1) was particularly significant, indicating that a part of the phenyl ring of MPT entered into the cavity of β-CD. (**b**) Curve-fitting analysis by the HypNMR2008 program.

The principle on which DOSY is based is very simple and can be summarized as follows: when the host and the guest are in the free state, they have their own diffusion coefficient that depends on their molecular weight and their shape. However, when they interact tightly together to form a complex, they behave as a single molecular entity and therefore should have the same diffusion coefficient [48,49].

Taking into account the fact that we are studying a rapid equilibrium on the NMR time scale between bound and free guest molecules, the observed (measured) diffusion coefficient (D_{obs}) is the weighted average of the free and bound diffusion coefficients (D_{free} and D_{bound}, respectively) and can therefore be used to calculate the bound fraction p by using the following Equation (1):

$$D_{obs} = p{\cdot}D_{bound} + (1-p){\cdot}D_{free} \tag{1}$$

which can be rearranged to yield:

$$p = \frac{D_{free} - D_{obs}}{D_{free} - D_{bound}} \tag{2}$$

where p is the fraction of complexed substrate molecules.

After binding of a small guest molecule (MPT) to a large host molecule (β-CD) the diffusion coefficient of the host is not greatly perturbed, therefore, the diffusion coefficient of the host–guest complex can be assumed to be the same as that of the non-complexed host molecule [50].

Pseudo 2D DOSY spectra are shown in Figure 5. The f1 dimension represents the self-diffusion coefficient (D) and the f2 dimension reports the chemical shift. The f1 is specific for each molecule, so the protons belonging to the same molecule appear in the same f1 row. The diffusion coefficients (D) and the fraction of complexed MPT molecules (p) measured at 23 °C in D_2O are reported in Table 2.

As expected, the D value of encapsulated MPT ($4.87{\cdot}10^{-6}$ $cm^2{\cdot}s^{-1}$) was lower than that of free MPT ($6.205{\cdot}10^{-6}$ $cm^2{\cdot}s^{-1}$) (Table 2, Figure 5) thus proving that MPT is included in the β-CD cavity and diffuses more slowly. Recalling that the association constant, K_a, for a 1:1 host–guest equilibrium of the type H + G ⇄ HG is defined by:

$$K_a = \frac{[HG]}{[H][G]} \tag{3}$$

where [H], [G], and [HG] are the equilibrium concentrations of the free host, free guest, and complex, respectively, Equation (3) can be rewritten as a function of the molar fraction [48] as:

$$K_a = \frac{p}{(1-p)([H]_0 - p[G]_0)} \quad (4)$$

where $[H_0]$ and $[G_0]$ are the total concentrations of the host and guest, respectively.

Lastly, inserting the value of the molar fraction just obtained from the DOSY experiments in the above Equation (4) we can calculate K_a by using the single-point procedure [51,52]. The association constant measured in this way was 654 M^{-1} at 23 °C (Table 2), and although the value obtained by the single-point approximation method results in large uncertainty, it is consistent with the corresponding value of 851 M^{-1} estimated via NMR titration.

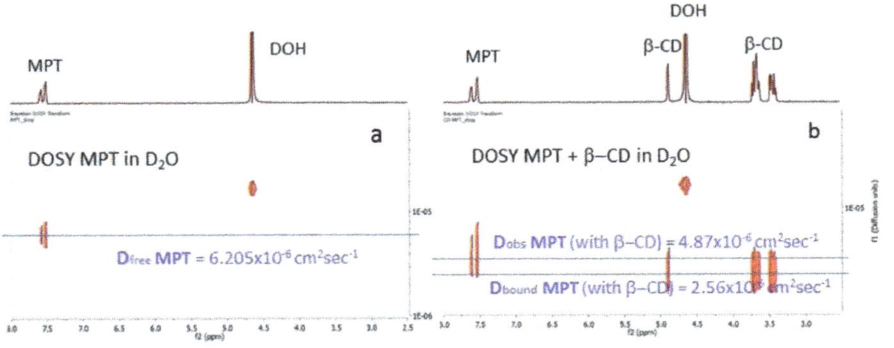

(**a**) (**b**)

Figure 5. Pseudo 2D DOSY spectra of MPT (**a**) 1.5 mM; (**b**) in the presence of β-CD 1.4 mM in D_2O, at 23 °C. (See Supplementary Information, SI, for details).

Table 2. Diffusion coefficients (D, 10^{-6} $cm^2{\cdot}s^{-1}$) related to the itemized protons of MPT_{free} (MPT alone 1.5 mM), $MPT_{(+\,\beta\text{-CD})}$ (MPT 1.5 mM with β-CD 1.4 mM), and $\beta\text{-CD}_{(+\,MPT)}$ (MPT 1.5 mM with β-CD 1.4 mM), fraction of complexed MPT and β-CD molecules (p) and association constant for the β-CD–MPT complex (K_a).

Chemical Species	Proton (ppm)	$D^{a,b}$	$D_{average}$	p_{bound}	K_a (M^{-1})
$MPT_{(free)}$	H_o (7.59)	6.22	6.205	0	–
	H_m/H_p (7.51) [c]	6.19	–	–	–
$MPT_{(+\,\beta\text{-CD})}$	H_o (7.62)	4.88	4.87	0.36	654
	H_m/H_p (7.52) [c]	4.86	–	–	–
$\beta\text{-CD}_{(+\,MPT)}$	H_m/H_p (7.52) [c]	2.256	2.56	0.36	–
	$H_3/H_5/H_6$ (3.67) [c]	2.57	–	–	–
	H_2/H_4 (3.43) [c]	2.55	–	–	–

[a] in D_2O at 23 °C; [b] estimated errors <5%; [c] overlapped signals.

3.2. ESI Mass Spectra

In order to provide further confirmation of the formation of the β-CD–MPT inclusion complex, some ESI-MS experiments were conducted in aqueous solutions containing MPT and β-CD in the ratio 1:1, 1:2, and 2:1. In all cases, a base peak at m/z 1311 corresponding to a 1:1 host–guest complex was detected (Figure 6).

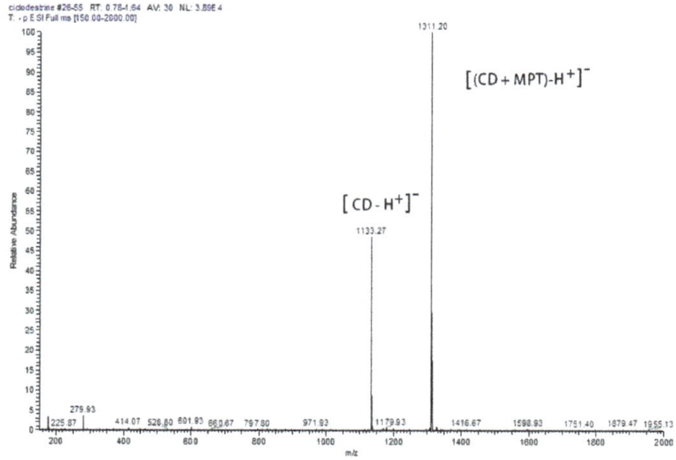

Figure 6. Electrospray negative-ion mass spectrum (ESI-MS) of a 1:1 β-CD–MPT aqueous solution that revealed a base peak corresponding to the 1:1 host–guest complex at m/z 1311 and a peak at m/z 1133 (50%) relative to uncomplexed β-CD.

3.3. FTIR Spectra

The diffuse reflectance FTIR spectrum of the solid β-CD–MPT complex precipitated from equimolar β-CD and MPT aqueous solutions was recorded and compared to those obtained on pure MPT and β-CD. Figure 7 clearly shows that the spectrum of the complex almost completely overlapped that of β-CD, but two extra peaks at 1492 cm^{-1} and 1593 cm^{-1} occurred (as evidenced in the enlarged inset), which corresponded to intense bands of the MPT molecule and did not appear in the β-CD spectrum. According to X.R. Ye et al. [53], both peaks were connected to the C–C stretching of the phenyl ring in MPT and the former also corresponded to N–H bending. These spectra further confirmed the complex formation.

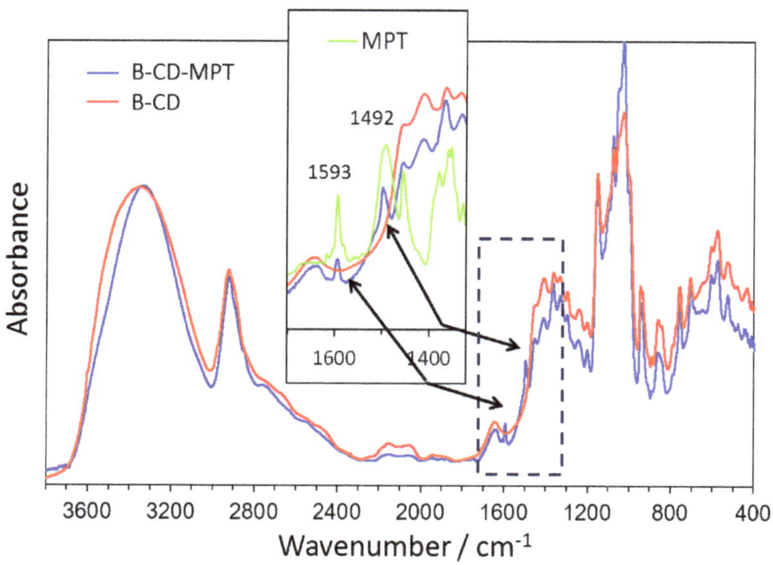

Figure 7. FTIR spectra of solid β-CD–MPT, MPT, and β-CD.

3.4. Electrochemical Tests

These tests were performed on dip coated bronze electrodes in ARX10, because the association of the obtained thin coatings and the higher aggressiveness of the concentrated solution allowed for the differentiation of the coating protectiveness within the 20 day immersion period.

Figure 8 collects the time evolution of R_p and E_{cor} values in this environment at 30 °C. For bare electrodes, high initial R_p values (9.1 kohm·cm^2) were obtained, which decreased quickly to about 1 kohm·cm^2 for immersions longer than 1 h and then increased again up to 4.6 kohm·cm^2 towards the end of the immersion period (Figure 8a). PropS-SH coatings showed much higher initial R_p values (almost 700 kohm·cm^2), which decreased by about 1 order of magnitude during the 20 days of immersion due to the slow penetration of the aggressive solution through the silane network. The addition of MPT to the silane solution determined rather low and constant R_p values, close to 20 kohm·cm^2. This behavior was not investigated but it is plausible that MPT interfered with the coating reticulation and/or a surface competitive adsorption between free MPT molecules on one side and the silanol and thiol groups of silane coatings on the other occurred, so impairing the coating adherence and performance. Instead, some improvements were achieved with the addition of β-CD, which due to its hydroxyl groups was likely capable of reacting with the silanol groups, so contributing to the silane network formation. Finally, a clear progression was observed after β-CD–MPT complex addition with high and rather constant R_p values (around 1 Mohm·cm^2). This suggests that beside the positive effect of β-CD, the release of MPT molecules from the β-CD cavity could also play an important role in corrosion inhibition.

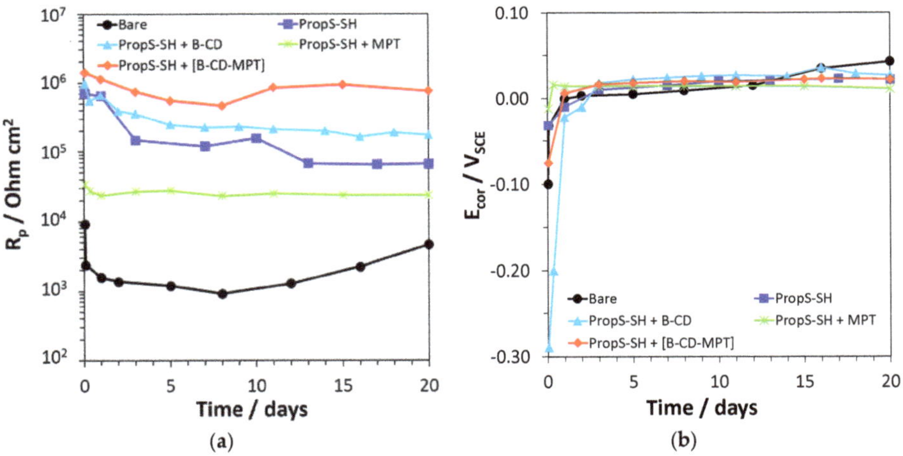

Figure 8. Polarization resistance (R_p) (**a**) and corrosion potentials (E_{cor}) values (**b**) measured during 20 days of immersion of bare and coated bronze electrodes in concentrated acid rain (ARX10) at 30 °C.

The E_{cor} values on bare electrodes evolved from −0.100 V_{SCE} after 1 h immersion to about +0.043 V_{SCE} after 20 days (Figure 8b). From previous research [54], it was found that this trend was the consequence of the degradation of the protective surface air-formed oxide film during the first half of the immersion period, which stimulated the cathodic reaction, with a consequent E_{cor} shift towards nobler values. Then, in the second half of the immersion, the progressive accumulation of surface corrosion products induced a slight inhibition of the anodic process and further consequent E_{cor} ennoblement. No significant E_{cor} differences were detected in the presence of the coatings. In all cases, E_{cor} increased to a certain extent during the initial 2 or 3 days of immersion, likely due to the evolution in the coating reticulation [39]. Then, they reached values in the range from 0.011 to 0.030 V_{SCE}, independently of the corresponding R_p values.

The polarization curves recorded at the end of the 20 days of immersion on bare and coated electrodes are shown in Figure 9, while Table 3 reports the electrochemical parameters derived from these curves.

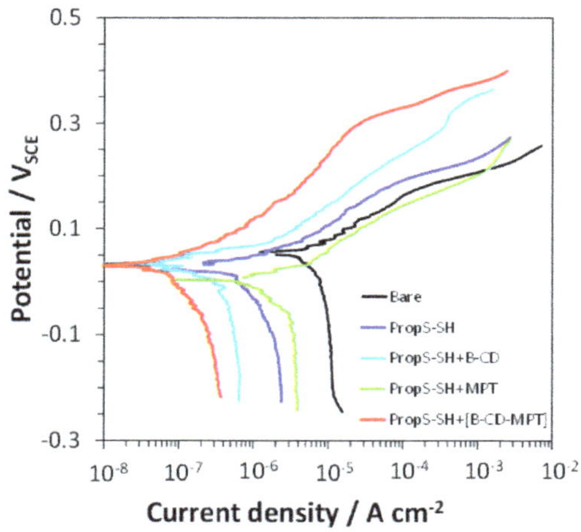

Figure 9. Polarization curves recorded on bare and silane (PropS-SH)-coated bronze specimens in the absence and in the presence of β-CD, MPT, or β-CD–MPT complex, after 20 days of immersion in ARX10 at 30 °C.

Table 3. Corrosion potentials (E_{cor}), corrosion current densities (i_{cor}), and anodic Tafel slopes (b_a) obtained on bare and coated bronze specimens after 20 days of immersion in ARX10 solution. Protection efficiencies (η) of silane coatings are also reported.

Coating Type	E_{cor}/V_{SCE}	$i_{cor}/\mu A \cdot cm^{-2}$	b_a/V	η/%
-	0.055	8.3	0.090	–
PropS-SH	0.024	1.3	0.092	84
PropS-SH + β-CD	0.030	0.70	0.088	92
PropS-SH + MPT	0.008	2.8	0.093	66
PropS-SH + β-CD-MPT	0.029	0.15	0.101	98

In particular, Table 3 collects the E_{cor} and corrosion current (i_{cor}) values, the anodic Tafel slopes, b_a, and the protection efficiency (η) of the coatings, evaluated by the formula:

$$\eta = \frac{i_{cor,b} - i_{cor,c}}{i_{cor,b}} \times 100 \quad (5)$$

where $i_{cor,b}$ and $i_{cor,c}$ are the corrosion currents evaluated on bare and coated electrodes, respectively. The cathodic Tafel slopes, b_c, which were not reported in Table 3, were generally much higher than b_a and close to infinity.

Figure 9 and Table 3 evidence that the coatings protect the underlying alloy from corrosion at different degrees. The plain PropS-SH coating mainly hindered the cathodic reaction, so determining a slight reactivation of E_{cor} in comparison to those of the bare electrodes, and afforded a final η of about 84%. In agreement with the R_p results, the addition of MPT to PropS-SH was detrimental to the coating protectiveness because it stimulated both the anodic and the cathodic reactions, suggesting a lower barrier effect of the coating and/or a lower surface adherence. In this case, the coating η value decreased down to 66% (Table 3). Conversely, the addition of β-CD or, even more, the β-CD–MPT complex in the coatings determined a decrease in the anodic and cathodic currents and induced η values of 92% and 98%, respectively. In contrast with MPT, the complex proved to be beneficial to the coating performances, likely due to a higher compatibility of the external β-CD surface with the silane network and to the specific inhibition afforded by released inhibitor molecules as evidenced by the following test.

3.5. Cyclic AR Spray Test

The self-healing capability of the silane coatings connected to the inhibitor release and adsorption at corrosion sites was evaluated during four weeks of exposure of the coated specimens with cross cut scratches to cyclic AR spray.

At the end of the test (Figure 10), the bronze coupons evidenced that the PropS-SH coating was rather protective at a distance from the scratches, but obviously did not avoid the substrate corrosion in scratched areas from which in fact the underfilm corrosion propagated. This corrosion form and coating delamination were less evident in the presence of β-CD, suggesting that this substance also increased the coating adherence, so improving the overall coating protectiveness. However, again, it did not prevent corrosion in the scratches. The addition of MPT in PropS-SH could not avoid the spread of corrosion attacks from the scratches and induced a significant surface color change (a brightening). Only β-CD–MPT complex addition significantly enhanced the substrate corrosion resistance and completely suppressed corrosion in the scratches and underfilm. This behavior suggested a self-healing capability of this coating type in the case of coating defects and mechanical damages, so prolonging the coating effectiveness.

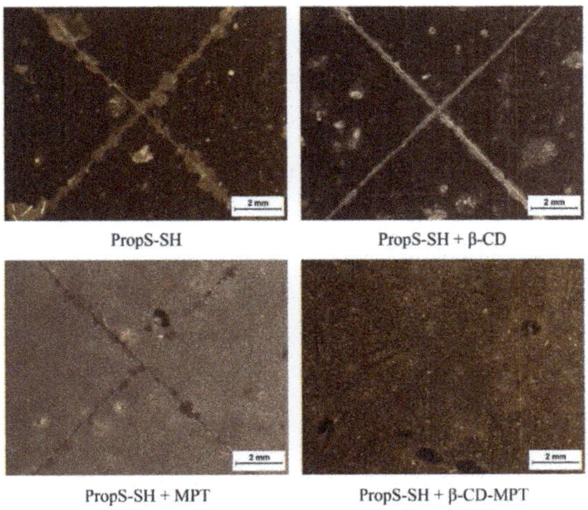

Figure 10. Surface aspect of cross-cut coupons at the end of a 4 week exposure to a cyclic acid rain (AR) spray test; only the PropS-SH coating containing the β-CD–MPT complex prevented the development of a corrosion attack starting from the scratches.

4. Conclusions

- The stability and the molecular structure of the host–guest β-CD–MPT complex was assessed by NMR, FTIR, and MS techniques.
- The analysis of the Complexation Induced Shifts suggested an inclusion structure of the complex, with partial insertion of the phenyl moiety of MPT at the wide rim of the hydrophobic β-CD cavity and partial protrusion of the tetrazole moiety of the inhibitor from the narrow rim of the host cavity.
- According to the DOSY experiments, the stability constant of the β-CD–MPT complex was 654 M^{-1}, in good agreement with the value of 851 M^{-1} obtained by curve fitting the NMR binding isotherm.

- The PropS-SH coating with entrapped β-CD–MPT complex exhibited an improved protectiveness (η = 98%) against bronze corrosion in comparison to plain PropS-SH (η = 84%) or PropS-SH containing only MPT (η = 66%) or β-CD (η = 92%).
- The PropS-SH coating with entrapped β-CD–MPT complex exhibited self-healing properties on bronze during exposures to cyclic AR spray at 35 °C.

Supplementary Materials: Supplementary materials are available online at http://www.mdpi.com/2079-6412/9/8/508/s1, Table S1: Parameters related to the titration, Figure S1: Variations of a portion of 1H NMR spectrum (400 MHz) of MPT 2.8 mM during its titration with β-CD 12 mM in D2O, the molar fractions of host are reported in Table S1, Figure S2: Stacked plot and computational analysis of the DOSY experiment of a sample containing MPT alone. MPT signal decays according to the gradients (G) together with the corresponding graphical analysis of the data. The diffusion coefficients are indicated with the letter F, Figure S3. Stacked plot and computational analysis of the DOSY experiment of a sample containing MPT + β-CD. MPT signal decays according to the gradients (G) together with the corresponding graphical analysis of the data. The diffusion coefficients are indicated with the letter F, Figure S4: Stacked plot and computational analysis of the DOSY experiment of a sample containing MPT + β-CD. β-CD signal decays according to the gradients (G) together with the corresponding graphical analysis of the data.

Author Contributions: Conceptualization and Methodology, C.M., G.F., and A.B.; Formal analysis C.M., G.F., and A.B.; Investigation G.F., G.D.C., C.M., F.Z., and A.B.; Writing—Review and Editing C.M., G.F., C.M., F.Z., and A.B.

Funding: This research was carried out in the scope of the B-IMPACT project (project ID 1149) of the M-ERA.NET 2013 call, supported by the national funding organizations of the partners Slovenia (MIZS), Italy (MIUR), and France (RMP).

Acknowledgments: The authors also gratefully acknowledge the University of Ferrara for financial support (FAR funds). Thanks are also given to P. Formaglio for NMR spectroscopic experiments and to T. Bernardi for mass spectrometric experiments.

Conflicts of Interest: The authors declare no conflict of interest.

References

1. Tasić, Ž.Z.; Petrović Mihajlović, M.B.; Radovanović, M.B.; Antonijević, M.M. New trends in corrosion protection of copper. *Chem. Pap.* **2019**, *73*, 2103–2132. [CrossRef]
2. Ashassi-Sorkhabi, H.; Moradi-Alavian, S.; Esrafili, M.D.; Kazempour, A. Hybrid sol-gel coatings based on silanes-amino acids for corrosion protection of AZ91 magnesium alloy: Electrochemical and DFT insights. *Prog. Org. Coat.* **2019**, *131*, 191–202. [CrossRef]
3. Zaferani, S.; Peikari, M.; Zaarei, D.; Danaee, I.; Fakhraei, J.; Mohammadi, M. Using silane films to produce an alternative for chromate conversion coatings. *Corrosion* **2013**, *69*, 372–387. [CrossRef]
4. Bierwagen, G.; Shedlosky, T.J.; Stanek, K. Developing and testing a new generation of protective coatings for outdoor bronze sculpture. *Prog. Org. Coat.* **2003**, *48*, 289–296. [CrossRef]
5. Lamaka, S.V.; Shchukin, D.G.; Andreeva, D.V.; Zheludkevich, M.L.; Möhwald, H.; Ferreira, M.G.S. Sol-gel/polyelectrolyte active corrosion protection system. *Adv. Funct. Mater.* **2008**, *18*, 3137–3147. [CrossRef]
6. Zhang, F.; Ju, P.; Pan, M.; Zhang, D.; Huang, Y.; Li, G.; Li, X. Self-healing mechanisms in smart protective coatings: A review. *Corros. Sci.* **2018**, *144*, 74–88. [CrossRef]
7. Montemor, M.F. Functional and smart coatings for corrosion protection: A review of recent advances. *Surf. Coat. Technol.* **2014**, *258*, 17–37. [CrossRef]
8. Shchukin, D.G.; Zheludkevich, M.; Yasakau, K.; Lamaka, S.; Ferreira, M.G.S.; Möhwald, H.; Shchukin, D. Layer-by-layer assembled nanocontainers for self-healing corrosion protection. *Adv. Mater.* **2006**, *18*, 1672–1678. [CrossRef]
9. Jafari, A.; Hosseini, S.; Jamalizadeh, E. Investigation of smart nanocapsules containing inhibitors for corrosion protection of copper. *Electrochim. Acta* **2010**, *55*, 9004–9009. [CrossRef]
10. Mahmoudian, M.; Nozad, E.; Kochameshki, M.G.; Enayati, M. Preparation and investigation of hybrid self-healing coatings containing linseed oil loaded nanocapsules, potassium ethyl xanthate and benzotriazole on copper surface. *Prog. Org. Coat.* **2018**, *120*, 167–178. [CrossRef]
11. Zheludkevich, M.L.; Poznyak, S.K.; Rodrigues, L.M.; Raps, D.; Hack, T.; Dick, L.F.; Nunes, T.; Ferreira, M.G.S. Active protection coatings with layered double hydroxide nanocontainers of corrosion inhibitor. *Corros. Sci.* **2010**, *52*, 602–611. [CrossRef]

12. Szejtli, J. Introduction and general overview of cyclodextrin chemistry. *Chem. Rev.* **1998**, *98*, 1743–1754. [CrossRef]
13. Hedges, A.R. Industrial applications of cyclodextrins. *Chem. Rev.* **1998**, *98*, 2035–2044. [CrossRef]
14. Uekama, K.; Hirayama, F.; Irie, T. Cyclodextrin drug carrier systems. *Chem. Rev.* **1998**, *98*, 2045–2076. [CrossRef]
15. Del Valle, E.M.M. Cyclodextrins and their uses: A review. *Process Biochem.* **2004**, *39*, 1033–1046. [CrossRef]
16. Li, S.; Purdy, W.C. Cyclodextrins and their applications in analytical chemistry. *Chem. Rev.* **1992**, *92*, 1457–1470. [CrossRef]
17. Takahashi, K. Organic reactions mediated by cyclodextrins. *Chem. Rev.* **1998**, *98*, 2013–2034. [CrossRef]
18. De Souza, T.M.; Cordeiro, R.F.; Viana, G.M.; Aguiar, L.C.; De Senna, L.F.; Malta, L.F.B.; D'Elia, E.; D'Elia, E. Inclusion compounds of dibenzylthiourea with hydroxypropylated-cyclodextrins for corrosion protection of carbon steel in acidic medium. *J. Mol. Struct.* **2016**, *1125*, 331–339. [CrossRef]
19. Casaletto, M.P.; Figà, V.; Privitera, A.; Mazzaglia, A.; Scala, A.; Zagami, R. Sustainable corrosion inhibition of Copper-based alloys by smart β-Cyclodextrin/Benzotriazole complexes. In Proceedings of the 5th European Cyclodextrin Conference, Lisbon, Portugal, 3–6 October 2017.
20. Khramov, A.; Voevodin, N.; Balbyshev, V.; Donley, M. Hybrid organo-ceramic corrosion protection coatings with encapsulated organic corrosion inhibitors. *Thin Solid Films* **2004**, *447*, 549–557. [CrossRef]
21. Khramov, A.N.; Voevodin, N.N.; Balbyshev, V.N.; Mantz, R.A. Sol–gel-derived corrosion-protective coatings with controllable release of incorporated organic corrosion inhibitors. *Thin Solid Films* **2006**, *514*, 174–181. [CrossRef]
22. Amiri, S.; Rahimi, A. Anticorrosion behavior of cyclodextrins/inhibitor nanocapsule-based self-healing coatings. *J. Coat. Technol. Res.* **2016**, *13*, 1095–1102. [CrossRef]
23. Altin, A.; Rohwerder, M.; Erbe, A. Cyclodextrins as carriers for organic corrosion inhibitors in organic coatings. *J. Electrochem. Soc.* **2017**, *164*, C128–C134. [CrossRef]
24. Varvara, S.; Bostan, R.; Bobis, O.; Găină, L.; Popa, F.; Mena, V.; Souto, R.M. Propolis as a green corrosion inhibitor for bronze in weakly acidic solution. *Appl. Surf. Sci.* **2017**, *426*, 1100–1112. [CrossRef]
25. Verma, C.; Ebenso, E.E.; Bahadur, I.; Quraishi, M. An overview on plant extracts as environmental sustainable and green corrosion inhibitors for metals and alloys in aggressive corrosive media. *J. Mol. Liq.* **2018**, *266*, 577–590. [CrossRef]
26. Marzorati, S.; Verotta, L.; Trasatti, S.P. Green corrosion inhibitors from natural sources and biomass wastes. *Molecules* **2019**, *24*, 48. [CrossRef]
27. Zucchi, F.; Trabanelli, G.; Fonsati, M. Tetrazole derivatives as corrosion inhibitors for copper in chloride solutions. *Corros. Sci.* **1996**, *38*, 2019–2029. [CrossRef]
28. Mihit, M.; Salghi, R.; Bazzi, L.; Hammouti, B.; Kertit, S.; El Issami, S.; Addi, E.A. A study of tetrazoles derivatives as corrosion inhibitors of copper in nitric acid. *Pigment Resin Technol.* **2006**, *35*, 151–157. [CrossRef]
29. Balbo, A.; Chiavari, C.; Martini, C.; Monticelli, C. Effectiveness of corrosion inhibitor films for the conservation of bronzes and gilded bronzes. *Corros. Sci.* **2012**, *59*, 204–212. [CrossRef]
30. Wolfe, J.; Grayburn, R. A review of the development and testing of Incralac lacquer. *J. Am. Inst. Conserv.* **2017**, *56*, 225–244. [CrossRef]
31. Aufray, M.; Balbo, A.; Benetti, F.; Bernardi, E.; Bignozzi, M.C.; Chiavari, C.; Esvan, J.; Gartner, N.; Grassi, V.; Josse, C.; et al. Protection of outdoor bronzes by eco-friendly and non-hazardous coatings based on silane and fluoropolymers: Results from the B-IMPACT project. In Proceedings of the Metal 2019, Neuchâtel, Switzerland, 2–6 September 2019; pp. 1–10.
32. Chiavari, C.; Balbo, A.; Bernardi, E.; Martini, C.; Bignozzi, M.; Abbottoni, M.; Monticelli, C.; Bignozzi, M. Protective silane treatment for patinated bronze exposed to simulated natural environments. *Mater. Chem. Phys.* **2013**, *141*, 502–511. [CrossRef]
33. Masi, G.; Balbo, A.; Esvan, J.; Monticelli, C.; Avila, J.; Robbiola, L.; Bernardi, E.; Bignozzi, M.; Asensio, M.; Martini, C.; et al. X-ray photoelectron spectroscopy as a tool to investigate silane-based coatings for the protection of outdoor bronze: The role of alloying elements. *Appl. Surf. Sci.* **2018**, *433*, 468–479. [CrossRef]
34. Chiavari, C.; Bernardi, E.; Balbo, A.; Monticelli, C.; Raffo, S.; Bignozzi, M.; Martini, C.; Bignozzi, M. Atmospheric corrosion of fire-gilded bronze: Corrosion and corrosion protection during accelerated ageing tests. *Corros. Sci.* **2015**, *100*, 435–447. [CrossRef]
35. Wu, D.H.; Chen, A.D.; Johnson, C.S., Jr. An improved diffusion-ordered spectroscopy experiment incorporating bipolar-gradient pulses. *J. Magn. Reson. Ser. A* **1995**, *115*, 260–264. [CrossRef]

36. Fielding, L. Determination of association constants (K_a) from solution NMR data. *Tetrahedron* **2000**, *56*, 6151–6170. [CrossRef]
37. Frassineti, C.; Ghelli, S.; Gans, P.; Sabatini, A.; Moruzzi, M.S.; Vacca, A. Nuclear magnetic resonance as a tool for determining protonation constants of natural polyprotic bases in solution. *Anal. Biochem.* **1995**, *231*, 374–382. [CrossRef]
38. Zucchi, F.; Frignani, A.; Grassi, V.; Trabanelli, G.; DalColle, M. The formation of a protective layer of 3-mercapto-propyl-trimethoxy-silane on copper. *Corros. Sci.* **2007**, *49*, 1570–1583. [CrossRef]
39. Chiavari, C.; Balbo, A.; Bernardi, E.; Martini, C.; Zanotto, F.; Vassura, I.; Bignozzi, M.; Monticelli, C.; Bignozzi, M. Organosilane coatings applied on bronze: Influence of UV radiation and thermal cycles on the protectiveness. *Prog. Org. Coat.* **2015**, *82*, 91–100. [CrossRef]
40. Masi, G.; Josse, C.; Esvan, J.; Chiavari, C.; Bernardi, E.; Martini, C.; Bignozzi, M.C.; Monticelli, C.; Zanotto, F.; Balbo, A.; et al. Evaluation of the protectiveness of an organosilane coating on patinated Cu-Si-Mn bronze for contemporary art. *Prog. Org. Coat.* **2019**, *127*, 286–299. [CrossRef]
41. Stern, M.; Geary, A.L. Electrochemical polarization I. A theoretical analysis of the shape of polarization curves. *J. Electrochem. Soc.* **1957**, *104*, 56–63. [CrossRef]
42. Pessine, F.B.T.; Calderini, A.; Alexandrino, G.L. Review: Cyclodextrin inclusion complexes probed by NMR techniques. *Magn. Reson. Spectr.* **2012**, *2012*, 237–264.
43. Kfoury, M.; Auezova, L.; Greige-Gerges, H.; Ruellan, S.; Fourmentin, S. Cyclodextrin, an efficient tool for trans-anethole encapsulation: Chromatographic, spectroscopic, thermal and structural studies. *Food Chem.* **2014**, *164*, 454–461. [CrossRef]
44. Schneider, H.-J.; Hacket, F.; Rüdiger, V.; Ikeda, H. NMR studies of cyclodextrins and cyclodextrin complexes. *Chem. Rev.* **1998**, *98*, 1755–1786. [CrossRef]
45. Bothner-By, A.A.; Stephens, R.L.; Lee, J.; Warren, C.D.; Jeanloz, R.W. ChemInform Abstract: Structure determination of a tetrasaccharide: transient nuclear overhauser effects in the rotating frame. *Chem. Inf.* **1984**, *15*, 811–813. [CrossRef]
46. Holm, R.; Østergaard, J.; Schőnbeck, C.; Jensen, H.; Shi, W.; Peters, G.H.; Westh, P. Determination of stability constants of tauro- and glycoconjugated bile salts with the negatively charged sulfobutylether-β-cyclodextrin: Comparison of affinity capillary electrophoresis and isothermal titration calorimetry and thermodynamic analysisof the interaction. *J. Incl. Phenom. Macrocycl. Chem.* **2014**, *78*, 185–194. [CrossRef]
47. Cameron, K.S.; Fielding, L. NMR diffusion coefficient study of steroid–cyclodextrin inclusion complexes. *Magn. Reson. Chem.* **2002**, *40*, S106–S109. [CrossRef]
48. Rymdén, R.; Carlfors, J.; Stilbs, P. Substrate binding to cyclodextrins in aqueous solution: A multicomponent self-diffusion study. *J. Incl. Phenom. Macrocycl. Chem.* **1983**, *1*, 159–167.
49. Mayzel, O.; Cohen, Y. Diffusion coefficients of macrocyclic complexes using the PGSE NMR technique: Determination of association constants. *J. Chem. Soc. Chem. Commun.* **1994**, *16*, 1901–1902. [CrossRef]
50. Wimmer, R.; Aachmann, F.L.; Larsen, K.L.; Petersen, S.B. NMR diffusion as a novel tool for measuring the association constant between cyclodextrin and guest molecules. *Carbohydr. Res.* **2002**, *337*, 841–849. [CrossRef]
51. Cameron, K.S.; Fielding, L. NMR diffusion spectroscopy as a measure of host–guest complex association constants and as a probe of complex size. *J. Org. Chem.* **2001**, *66*, 6891–6895. [CrossRef]
52. Šmejkalová, D.; Piccolo, A. Host-guest interactions between 2,4-dichlorophenol and humic substances as evaluated by ^{1}H NMR relaxation and diffusion ordered spectroscopy. *Environ. Sci. Technol.* **2008**, *42*, 8440–8445. [CrossRef]
53. Ye, X.; Xin, X.; Zhu, J.; Xue, Z. Coordination compound films of 1-phenyl-5-mercaptotetrazole on copper surface. *Appl. Surf. Sci.* **1998**, *135*, 307–317. [CrossRef]
54. Monticelli, C.; Balbo, A.; Esvan, J.; Chiavari, C.; Martini, C.; Zanotto, F.; Marvelli, L.; Robbiola, L. Evaluation of 2-(salicylideneimino) thiophenol and other Schiff bases as bronze corrosion inhibitors by electrochemical techniques and surface analysis. *Corros. Sci.* **2019**, *148*, 144–158. [CrossRef]

coatings

MDPI

Article

Influence of Growth Defects on the Corrosion Resistance of Sputter-Deposited TiAlN Hard Coatings

Peter Panjan [1,*], Aljaž Drnovšek [1], Peter Gselman [1,2], Miha Čekada [1], Matjaž Panjan [1], Tonica Bončina [3] and Darja Kek Merl [1,†]

1 Department of Thin Films and Surfaces, Jožef Stefan Institute, Jamova cesta 39, SI-1000 Ljubljana, Slovenia
2 Interkorn d.o.o., Gančani 94, SI-9231 Beltinci, Slovenia
3 Faculty of Mechanical Engineering, University of Maribor, Smetanova 17, SI-2000 Maribor, Slovenia
* Correspondence: peter.panjan@ijs.si; Tel.: +386-1-477-3278
† Dr. Darja Kek Merl passed away before publication was completed.

Received: 10 July 2019; Accepted: 9 August 2019; Published: 12 August 2019

Abstract: In this work, the causes of porosity of TiAlN hard coatings sputter deposited on D2 tool steel were studied since its corrosion resistance is mainly affected by imperfections within the coating (e.g., pinholes, pores, crevices). The corrosion test was performed in a chlorine solution using electrochemical impedance spectroscopy. The coating morphology of growth defects before and after the exposure was studied by scanning electron microscopy (SEM), while focused ion beam (FIB) was used to make series of cross-sections through individual selected defects. We confirm that pitting corrosion is closely related to the through-thickness growth defects. It was also found that in the case of nodular defects, the intensity of corrosion depends on the shape of the seed.

Keywords: PVD hard coatings; growth defects; pinhole; focused ion beam; pitting corrosion

1. Introduction

Hard coatings prepared by physical vapor deposition (PVD) are widely used to improve the wear resistance of all types of tools, to reduce the friction of machine parts, for decorative applications, and for protection of implants and medical instruments [1–3]. In the majority of these applications, the corrosion attack of tools and components occurs by the corrosive environment to which they are exposed (e.g., corrosive gases, technical reagents, cooling and lubricant liquids, body fluids, sea water). In general, such corrosion attacks drastically reduce the lifetime of coated tools and components.

Although the PVD hard coatings have a very good corrosion resistance, they do not provide good corrosion protection of the less noble substrate materials [4,5]. The corrosive attack of coated components is caused by imperfections within the coating [3,4,6–13]. It is well known that due to the line-of-sight transfer of the vapor flux during the sputtering process, many intrinsic defects are generated in the coating [14]. Morphological imperfections caused by the shadowing effect include columnar structures, pinholes, pores, void zones around nodular defects or droplets (crevices), microcracks, grain boundaries, and other discontinuities. All these microstructure imperfections can significantly affect their corrosion resistance, if they extend throughout the whole thickness of the coating. When the coated substrate is exposed to a corrosive medium, the electrolyte penetrates to the substrate through such defects, driven by the capillary forces. In the case of a less noble substrate material, this leads to the formation of local galvanic corrosion between the substrate (acting as an anode) and the coating (acting as a cathode) [5]. The corrosion reaction takes place under a closed electrical circuit in which the current flow is based on the electrical (electron) conductivity of the coating on one hand and on the ionic conductivity of the electrolyte (it is conductive due to the presence of dissolved cations and anions) on the other hand [15]. Corrosion is localized to the defect area, where the anodic dissolution of the exposed substrate material appears. The corrosion attack is more intense if the difference of the electrochemical potential between the coating and the substrate in the selected electrolyte is larger. Since the area of the cathode (coating) is much larger than that of the anode (substrate), pitting corrosion is the predominant corrosion type. For the same reason, the corrosion process in the pores and pinholes is fast. Pits which are present in the substrate material are extending radially from the pores, resulting in cracking and removal of the coating above by flaking (blistering). In such a way, pitting corrosion has a devastating effect on coated tools or components.

Many ways to improve the corrosion resistance are described in the literature [4,5,16]. In general, any way which reduces the porosity of the coating improves the corrosion resistance at the same time. Although it is practically impossible to eliminate the coating defects completely, their density can be reduced by proper pretreatment of the substrate surface, coating design, and modification of the deposition process. A lot of research has been devoted to this problem in the past. It was found that the substrate surface roughness must be as low as possible since the higher roughness results in less complete coverage of the substrate with the coating material (due to the shadowing effects during the PVD coating growth). It was also found that the corrosion resistance increases with increasing coating thickness due to reduced mass transport between the bottom of the pinhole and the corrosive environment. Jehn estimated that a satisfactory corrosion resistance of the coating can be achieved only if its thickness is more than 14 μm [4].

More recently, the sealing of the PVD hard coatings with only a few-nanometers-thick (oxide) film prepared by low-temperature atomic layer deposition (ALD) technique was introduced [17,18]. This technique offers the deposition of uniform, conformal, and high-density thin films with a thickness of only a few nanometers. Such a thin layer (e.g., Al_2O_3) blocks pinholes and other defects in the PVD hard coating.

The next approach for improvement of corrosion resistance is the addition of a sacrificial element (e.g., Mg) to conventional hard coatings [19].

In the past, efforts were made to improve the corrosion resistance of PVD hard coatings by appropriate coating architecture (deposition of interlayers, multilayer coatings, duplex coatings) [20–22]. Using these approaches, the improvement of corrosion performance is the result of the synergistic effect of two or more materials.

The corrosion behavior of PVD hard coatings also depends on the coating microstructure. In general, PVD hard coatings designed for corrosion protection should possess a dense and uniform microstructure (fine-grained or amorphous). The columnar microstructure is not desirable because the interfaces between adjacent columns can act as pathways for corrosive media to reach the substrate.

The growth of the columnar microstructure of the PVD hard coating can be interrupted by intermediate ion etching of the growing coating [23,24]. It is assumed that ion etching leads to a denser coating, due to defect-induced renucleation and improved adatom mobility. New nucleation sites for subsequent film growth are also generated and a fine-grained microstructure is formed. In such a way, the path length for corrosive medium is increased. By ion etching, the material from grain flanks is redeposited to the pore walls resulting in a decrease of porosity.

Nitride hard coatings lose their columnar microstructure and become denser if doped by oxygen [25]. Additionally, with increasing oxygen content, the electrical conductivity considerably decreases. Therefore, the iron dissolving is strongly reduced because the electrons cannot travel through the coating.

Another good approach to obtain a dense microstructure of the coatings and to overcome intrinsic defects is to add other elements (Si or B) into the hard coatings to form nanocomposite coatings with nanosized crystallites surrounded by the matrix [23].

Yet another new strategy is based on high-power impulse magnetron sputtering (HiPIMS), which allows an energetic condensation [24–27]. Such PVD coatings have a higher density, better adhesion, and smooth surface (free of macroparticles), while pinholes can be closed by higher adatom mobility. An additional advantage is etching with metal ions produced by the HiPIMS plasma. The metal ions do not only etch the substrates, they are also implanted in the near-surface layer of the substrate.

Such metal-ion-implanted interlayer is normally beneficial since it improves adhesion and corrosion resistance of the coating.

While the mechanism of pitting corrosion of coatings is quite well described [4,7,9], we know only a little about the formation of microstructure imperfections and their influence on the corrosion resistance of hard coatings. Previously, we tried to explain the origin of the growth defects in PVD hard coatings [28,29]. In this paper, we focused on the identification of those growth defects in sputter-deposited TiAlN hard coatings, where the pitting corrosion occurs after it has been exposed to the corrosive medium.

2. Experiment

The industrial magnetron sputtering system (CC800/7, CemeCon, Wurselen, Germany) was used for deposition of the TiAlN hard coating. The thickness of the coatings was around 4.2 µm, using a twofold planetary rotation. The samples analyzed in this study consisted of test plates made of D2 tool steel. They were first ground and polished up to the roughness of about 10 nm (R_a). Before deposition, they were cleaned in detergents and ultrasound, rinsed in deionized water, and dried in hot air. In the vacuum chamber, they were first heated to about 450 °C, then cleaned in situ by ion etching. Standard TiAlN coating was deposited by DC sputtering of four segmental Ti–Al targets at 8 kW each. The working pressure was 750 mPa (N2, Ar, and Kr atmosphere), and the bias voltage was −100 V. After two-thirds of the deposition time, the deposition process was interrupted for an intermediate plasma etching process. The intermediate etching provides fine-grained and less porous microstructure (less pronounced columnar microstructure).

A corrosion test of TiAlN deposited on D2 tool steel, using electrochemical impedance spectroscopy (EIS), was performed in 0.5 M NaCl solution at pH = 3.8 using a Parstat 2263 spectroscope (Princeton Applied Research, Oak Ridge, United States of America). The test specimen employed as a working electrode was exposed to the solution for 96 h.

The coating microstructure and defect morphology, before and after the exposure, were studied in a planar surface view and in focused ion beam (FIB) cross-sectional view. We used the FIB integrated into the Helios Nanolab 650 electron microscope (FEI BV; now FEI SAS, part of Thermo Fisher Scientific, Merignac, France). The primary opening, required for further observations of cross-sections, with dimensions 22 × 12 × 12 μm^3, was ion-milled at 20 nA beam current, while the acceleration voltage of ions was 30 kV. The sections of the material were removed from the specimen surface by sputtering in three steps: rough removal (30 keV/65 nA), medium polishing (30 keV/20 nA), and fine polishing (30 keV/1 nA). After all three steps were completed, the image of the cross-section was taken using 10 pA ion beam current. After imaging, the next slice of the sample with a thickness of about 0.9 µm was removed by ion milling and another image of the new cross-section was acquired. This procedure was repeated until the entire site of interest was examined.

3. Results

3.1. Mechanisms of Pinhole Formation

As discussed in the introduction, one major drawback of PVD hard coatings is that they are not free from growth defects. Besides nodular and flake defects, open and closed pores, pinholes have been observed on the sputter-deposited TiAlN hard coating. All these defects have different shapes and sizes. However, pitting corrosion does not occur on all these sites. The question is how to identify which defects are responsible for pitting corrosion. It can only be those defects that are extending through the coating down to the coating–substrate interface. At least four groups of growth defects which can cause the pitting corrosion can be identified. The causes for their formation are discussed in more detail in the following text.

3.1.1. Pinhole Formation at Pits in the Substrate

During grinding and polishing of tool steel substrates, shallow protrusions are formed at the carbides because they are harder than the ferrous matrix [30]. Because of the high shear stress during polishing, a part of protruded carbides can be torn out, leaving a pit in the substrate (Figure 1a). Due to the line-of-sight nature of the PVD deposition process, a part of the pit (hole) wall remains uncovered, while a pinhole is formed in the coating (Figure 1b). Whether a pinhole will be formed or not depends on the pit depth-to-width ratio. In the case of high aspect ratio (a narrow but deep pit), the deposition starts to coat the upper sidewalls and corners of the feature, which shadows the lower area from subsequent deposition. During the coating growth, the hole's opening narrows, and if the PVD coating is thick enough, coatings from both sides of the hole touch and close the opening, forming an isolated pore (keyhole). However, a discontinuity of microstructure is preserved through the growing coating, extending up to the coating surface. The pinhole will not be able to close up if the width of the pit is comparable to the final coating thickness. In the case of a low aspect ratio (a laterally large but shallow pit), an open pinhole will be formed.

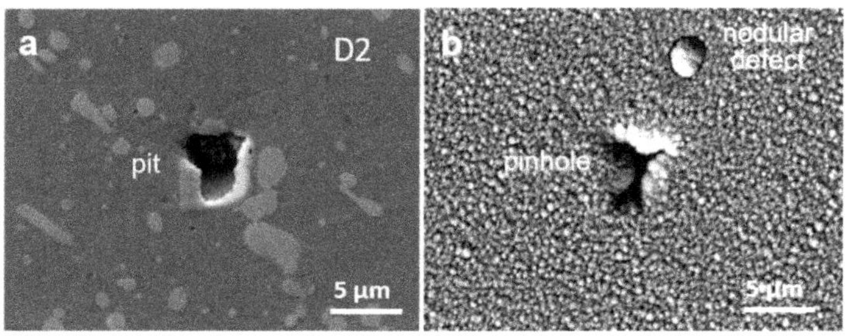

Figure 1. Pit on the D2 tool steel substrate surface at the site where a carbide grain was torn out from the surface of the D2 tool steel during the polishing process (**a**) and the same substrate surface area after deposition of a hard coating (**b**). Close to the pinhole, a nodular defect was also formed.

3.1.2. Pinhole Formation at Carbide and Nonmetallic Inclusions

During polishing of tool steel substrates, shallow protrusions are formed at the inclusions which are harder than the ferrous matrix (e.g., carbides and oxides), while shallow craters are formed at the inclusions which are softer (e.g., MnS) [29]. The reason is the difference in the removal rates of different inclusions in comparison to the ferrous matrix. Additional topographical changes occur during ion etching due to different etching rates of various phases in the tool steel material. Thus, the total geometrical extension (up or down) of the surface irregularities from the ferrous matrix level depends on the differences in both the removal rate and ion etching rate. Consequently, shallow craters (e.g., on the sites of Cr_7C_3, MnS) and protrusions (e.g., oxides) are formed. During the deposition process, all morphological features formed during the previous steps are transferred through the coating. Due to the shadowing effect, pinholes can be formed at positions of shallow craters (Figure 2). Formation of such a type of pinhole is displayed in Figure 2 for Al_2O_3/TiAlN double-layer hard coating sputter deposited on powder metallurgical (PM) tool steel substrate ASP30 in the CC800/9 deposition system. The bottom TiAlN layer was made in a similar way to that in CC800/7 (CemeCon, Wurselen, Germany), while the oxide top layer was prepared in the oxygen atmosphere by magnetron sputtering, operated in pulsed mode.

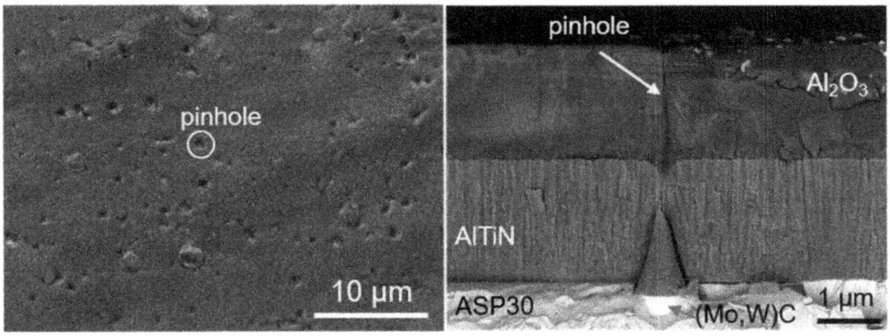

(**a**) (**b**)

Figure 2. Planar SEM view of a pinhole in the Al_2O_3/TiAlN hard coating deposited on ASP30 tool steel (**a**) and fracture cross-sectional SEM image of selected pinholes which appeared at shallow craters formed during ion etching at (W, Mo)C carbide in ASP30 tool steel substrate (**b**).

One of the origins of pinhole formation is also nonmetallic inclusions (e.g., MnS, Al_2O_3) in steel substrates. Such inclusions are formed as a result of the steelmaking process and are present in all types of steels. The removal rates (during polishing) and etching rates (during ion etching) of the inclusions are different in comparison with the ferrous matrix. In the case of MnS inclusions, for example, both rates are larger and therefore shallow craters are formed after both substrate pretreatments (Figure 3). Consequently, during the deposition process, such a shallow crater can be a starting point for a pinhole formation. Due to the generally poor contact between the inclusion surface and hard coating, a strong columnar growth with an extensive mesh of pinholes was observed. In the case of an Al_2O_3 inclusion with a complex microstructure (Figure 4a), a series of pinholes (Figure 4b) has formed on its boundary with the ferrous matrix.

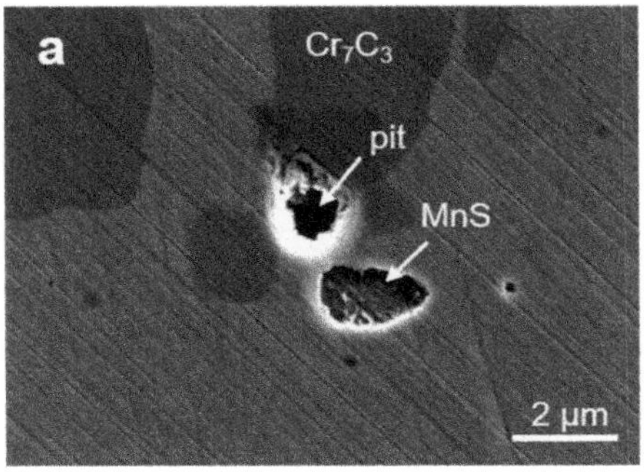

Figure 3. *Cont.*

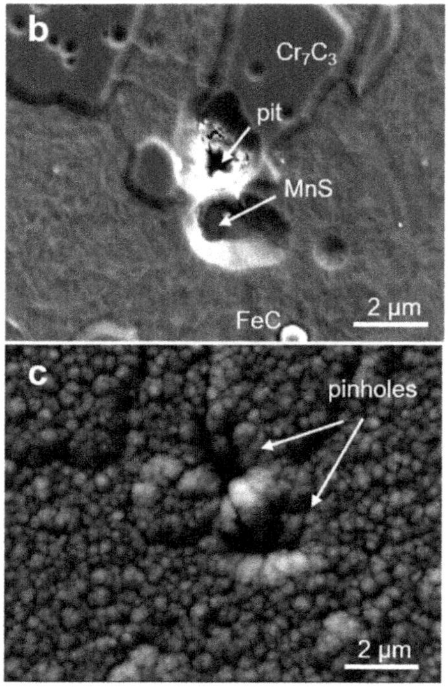

Figure 3. SEM images of the same MnS inclusion in the D2 substrate: (**a**) after polishing; (**b**) after ion etching; (**c**) after deposition of a hard coating. Above the MnS inclusion, a pit can be also observed on the polished substrate. In a similar way to the example in Figure 1, it caused the formation of the pinhole (**c**).

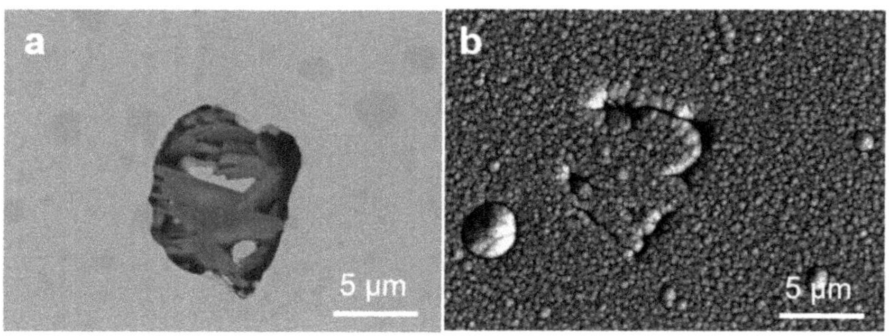

Figure 4. SEM images of the same Al_2O_3 inclusion in the D2 substrate after ion etching (**a**) and after deposition of hard coating (**b**).

3.1.3. Pinhole Formation at Nodular Defects

Even at the sites where nodular (Figures 5 and 6) and flake defects started to grow on the substrate surface, some degree of porosity remained. Nodular and flake defects are formed during deposition of PVD hard coating on the seed particles, which are either foreign or flake particles attached to the substrate surface before starting the deposition process. In the case of a nodular defect, shown in Figure 5a, the seed particle is based on iron. A step-like feature on the substrate surface is clearly visible (Figure 5b). The appearance of such a step indicates that the seed already existed on the substrate surface before ion etching. Therefore, this area was not exposed for etching. In general, the contact of the seed particle with the substrate is very poor, while the nodular defect is more or less separated from the coating matrix (Figure 5b). The border between the nodular defect and the undisturbed coating is therefore composed of a series of pinholes. The corrosion behavior of a hard coating does not only depend on the defect size, but also on the shape of the seed. Balzer et al. reported that the shape of the seed strongly affects the defect growth and the porosity of the coating matrix–defect boundary [31]. For a through-coating nodular and flake defect, the corrosion attack is initiated by galvanic or crevice corrosion. The void zones around nodular defects allow the solution penetration down to the steel substrate surface. If the formation of nodular defects is caused by droplets, then galvanic corrosion caused by differences of binding energy and chemical composition between the coating matrix and the droplet can also appear [6].

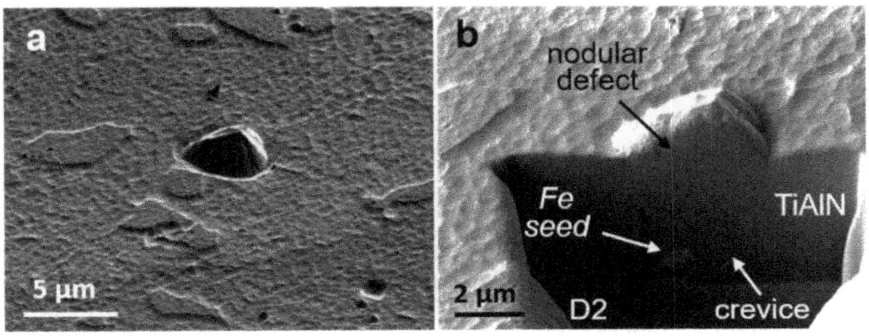

Figure 5. SEM planar view (**a**) and FIB cross-sectional image of a nodular defect (**b**) in the TiAlN hard coating sputter deposited on D2 steel.

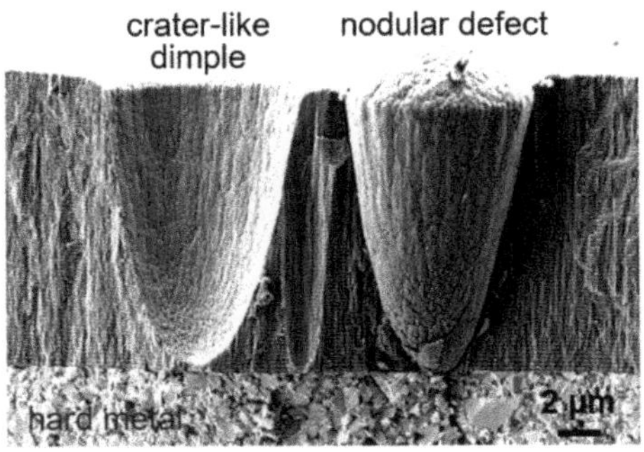

Figure 6. Fracture cross-sectional SEM image of the crater-like dimple which was left by the detachment of a nodular defect (**a**) and a nodular defect which has remained in the coating (**b**).

3.1.4. Pinhole Formation on Sites Where Nodular Defects Wrench out of the Coating

Another manifestation of defects is important in terms of corrosion resistance. This is a case when it comes to a detachment of the overgrown seed particles or droplets during or after deposition which leaves a crater-like dimple (Figure 6) behind. As already mentioned, this phenomenon happens because the bond between the seed particle and the nodular or flake defects with the surrounding coating matrix is poor. The whole defect or a part of it can be detached from the coating when residual stresses overcome the adhesion. The detachment probability increases with the coating thickness and the size of the nodular defect. Formation of such a hole may also be caused by external forces (such as cleaning of the coated sample by ultrasonic cavitation or wiping). If the nodular and flake defects are removed from the coating surface during the deposition process, then the crater is covered by the additional growing coating which is rather porous. If the nodular defect leaves the coating surface during the service of the coated component, a naked substrate area may be found at the bottom of these craters. If the craters, which are left by the detached nodules or flakes, are extending down to the substrate, then they offer an instant and easy access for the corrosive medium.

3.2. Electrochemical Impedance Measurement

A corrosion test of TiAlN deposited on D2 tool steel was performed in chloride medium using electrochemical impedance spectroscopy [32]. The exposition time was 96 h. The corrosion potential and polarization resistance (R_p) during the immersion in the corrosive medium can provide information regarding the ongoing corrosion reaction processes. Figure 7 shows the impedance response of the TiAlN/D2 coating systems in a Nyquist representation at different immersion times. Such diagrams offer some information regarding the performance of coatings deposited on less noble metal substrates. Thus, an initially high impedance (low corrosion rate) decreases with immersion time (this means that the corrosion rate increases). Higher corrosion rate can be explained by the fact that the coating defects open out, exposing more surface area of substrate to the NaCl solution. At low frequency, the interception with the real axis is ascribed to the polarization resistance (R_p, pure resistance at low frequency) at the corrosion potential. R_p resistance is the sum of resistances of coating, electrolyte in the pores, and charge–transfer (corrosion) reaction at the substrate/electrolyte interface. As the immersion time increases, the R_p values decrease for one order of magnitude, which indicates that the response becomes more conductive; indeed, with increasing time, the open voids and pinholes permit the solution to penetrate through to the steel substrate, causing localized corrosion. There are significant variations of corrosion current density, corrosion potential, and polarization resistance for nominally identical samples. This is consistent with corrosion caused by defects that are randomly distributed across the sample.

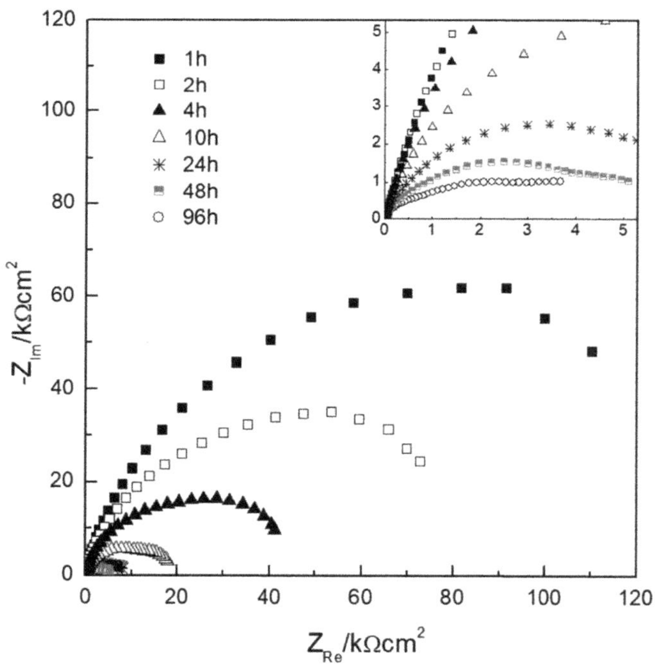

Figure 7. The impedance response of the TiAlN/D2 coating systems in a Nyquist representation at different immersion times.

3.3. *Typical Pitting Corrosion Attack at Growth Defects*

The inspection of the corroded sample was carried out in an optical microscope and SEM, following the electrochemical test. The corrosion attack under the coating, where the defect had been generated, was analyzed at the cross-section, prepared by the FIB technique. The inspection by optical and SEM microscope of the corroded sample revealed that pitting corrosion occurred only at certain growth defects. The measure of pitting corrosion intensity could be the area of the circular ring with corrosion products around the growth defect. The larger the radius, the more intensive pitting corrosion has been. However, the corrosion intensity of selected growth defects is not constant during the exposure to the corrosive media. In some cases, a certain time is necessary for pit formation, while later, the corrosion initiation sites are often filled with corrosion products (sealing effect) [5,13]. Corrosion products which accumulate at the pit sites hinder the local access to the electrolyte and thus reduce the corrosion kinetics.

The consecutive serial sectioning of thin slices performed by FIB technique and imaging of cross-sections gives us the insight into the defect internal structure, while EDX map analysis enables the determination of the composition of both seed as well as corrosion products. From the FIB cross-section, we also obtained the information on the intensity of the corrosion attack under the coating where the defect had been generated. A measure of the pitting corrosion intensity at a selected pinhole can also be the volume of the substrate material which had been removed.

We found that during the electrochemical test, different local defects (pinhole, pore, and crevice) caused localized corrosion of different intensities. Even in the case of the same type of defects, the corrosion intensity was very different, depending on how the pathways around the defect toward the substrate surface were open. In the following text, some typical examples are presented. Figure 8 shows the nodular defect, which started to grow on the substrate surface on an iron-based seed particle, but no traces of corrosion were observed. It is evident that the seed has a simple shape and smooth surface, while the boundary between the nodular defect and the coating matrix is mainly closed. It is completely different in the cases of nodular defects shown in Figures 9 and 10, where severe corrosion attacks occurred. Traces of corrosion products are still visible around the nodular defects, although the majority of the accumulated corrosion products were removed after the corrosion test by the sample-cleaning procedure. In the last two cases, the shape of seeds is more complex; they are composed from iron and chromium, indicating they are wear debris originating from the fixture system. In the case of severe corrosion (Figures 9 and 10), a crevice between the nodule and the coating matrix can be observed. Its width is not uniform and depends on the dimension and shape of the seed particle. This is consistent with the findings of Balzer et al. that the more complex seed geometry causes a more complex shading and the boundary between the nodular defect and the coating matrix is more porous [31]. The arrow in Figure 10c indicates a crevice where EDX maps (Figure 10) show traces of oxide products and chlorine, which proves that it was a pathway for solution penetration through the void zones around the nodular defects.

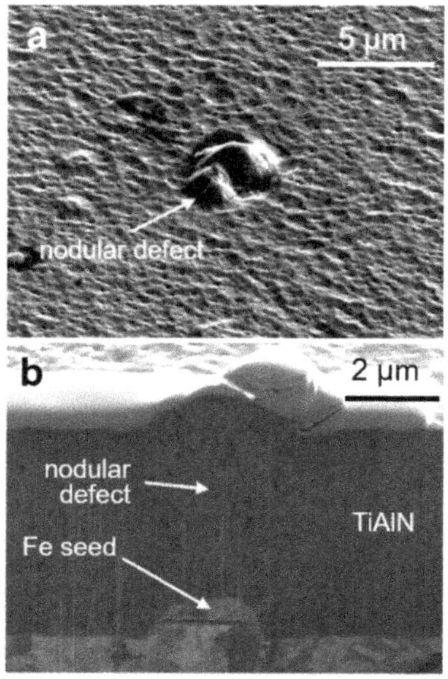

Figure 8. SEM top view (**a**) and FIB image (**b**) of a nodular defect where no pitting corrosion was observed. The nodular defect is intact and no crevices are visible at the boundary between the defect and the coating matrix.

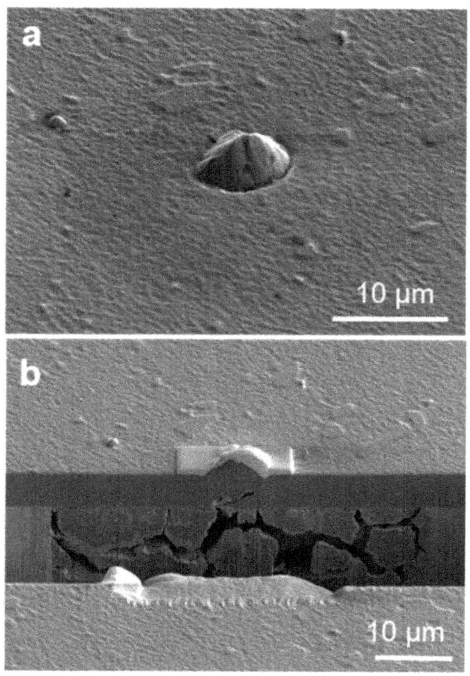

Figure 9. SEM top view (**a**) and FIB image (**b**) of a nodular defect where intensive pitting corrosion occurred.

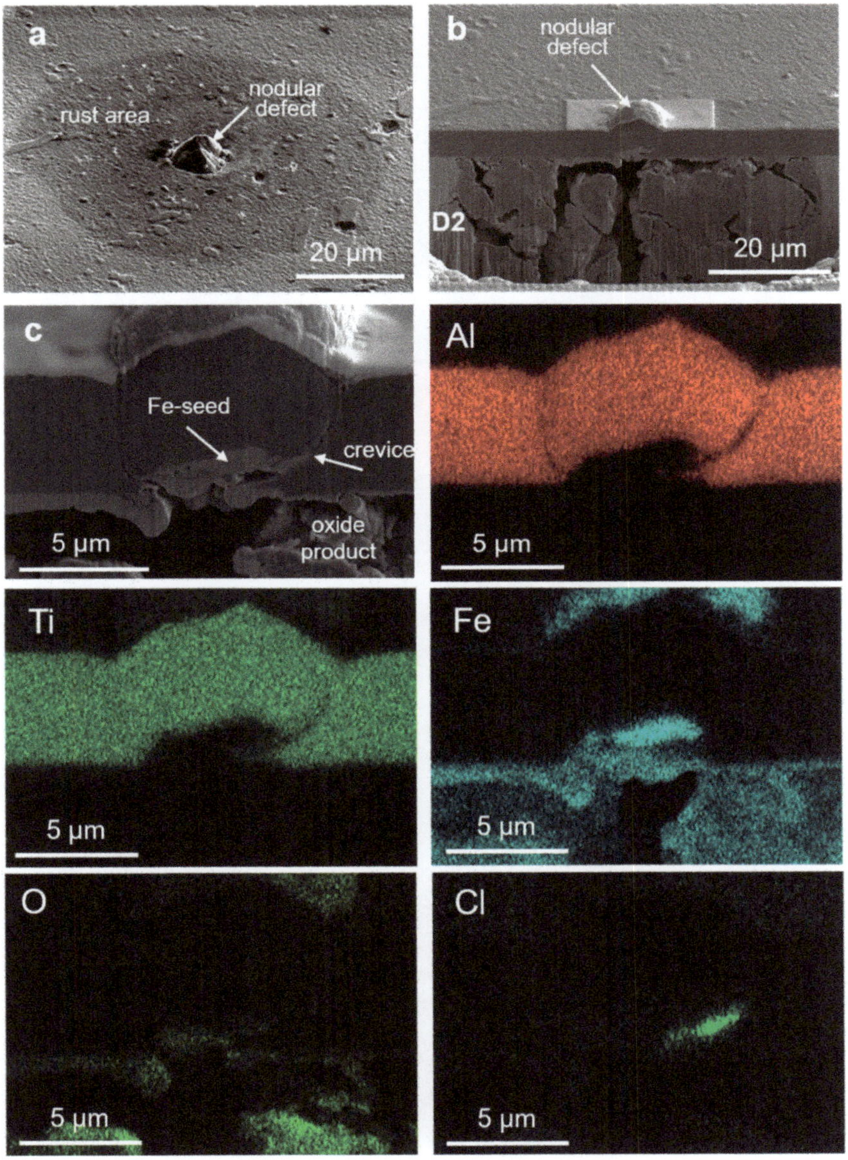

Figure 10. SEM top view (**a**) and FIB images of nodular defects and corresponding EDX maps where intensive pitting corrosion occurred (**b**,**c**).

Figure 11a shows another example where two defects in the TiAlN coating (pinhole, dish-like crater) were analyzed by the FIB. The right dish-like crater is much larger, but it is shallow and does not extend down to the substrate (Figure 11b). Because it is limited to the topmost part of the coating, there are no traces of corrosion at this site. On the other side, a very narrow pinhole in the coating on the left shows extensive damage in the substrate which is located underneath (Figure 11c). We can see that the site where the pinhole started is a pit at which a small piece of carbide grain was torn out, probably during mechanical pretreatment of the substrate surface. The similar type of pinhole in Figure 12a, generated on the same coated sample, caused a less intensive corrosion attack (Figure 12b) during the electrochemical test than that shown in Figure 11c, because the opening of the pinhole is evidently smaller.

The last case is the corrosion attack at a crater-like dimple formed on the site where nodular defects were detached (Figure 13). An extremely strong corrosion attack occurred at such kind of coating defect. The strength of the driving force for dissolution of steel in NaCl solution is seen from the volume of the cavity which was generated by the dissolved steel underneath the coating. Namely, all this dissolved material has been transferred through the small channels around the nodular defect.

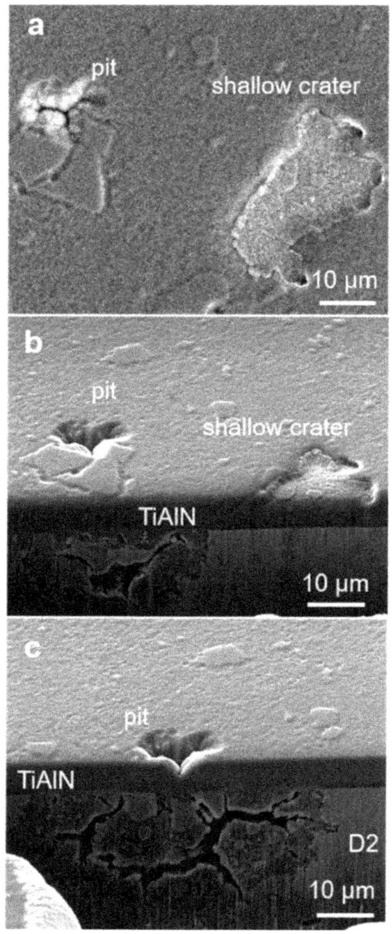

Figure 11. SEM top view of a pinhole and shallow crater (**a**) and FIB section across the crater (**b**) and pinhole center (**c**).

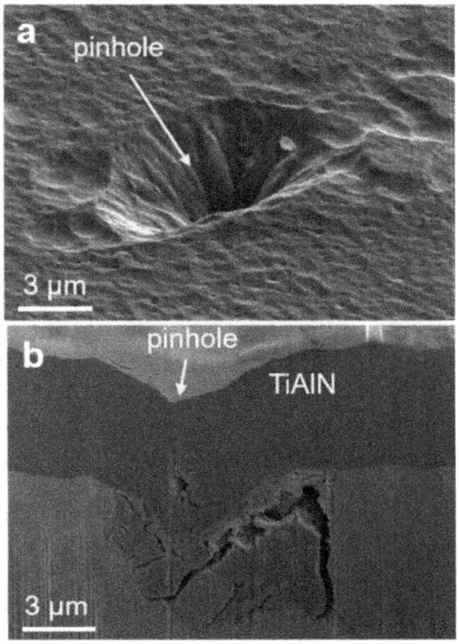

Figure 12. SEM top view (**a**) and FIB images of pinhole (**b**) where weak pitting corrosion occurs.

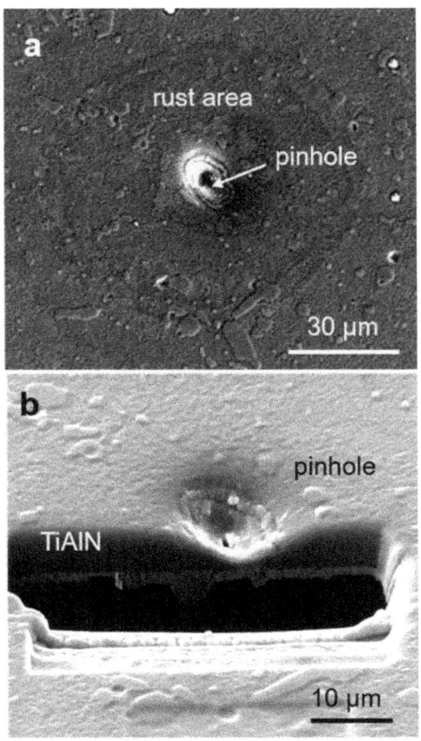

Figure 13. SEM top view of crater-like dimple (**a**) in sputter deposited D2 hard coating where extremly intensive pitting corrosion occured (**b**).

4. Conclusions

All PVD hard coatings contain some degree of microporosity (i.e., open pores, pinholes, void zones around the nodular defects or droplets, microcracks, and pores between the columns). In this study, we confirmed the deleterious effect of through-thickness growth defects on corrosion resistance of sputter-deposited TiAlN hard coating. We performed a systematic overview of all the mechanisms that lead to the creation of such defects. There are many possible causes of such defects: pits on the substrate surface formed during its pretreatment, different nonmetallic inclusions in the tool steels, foreign and flake particles which arrived on the substrate surface before the deposition process, localized delamination of coating, and microcracks. To prove the existence of such defects in the hard coating, we used the FIB technique. After the corrosion test performed in chlorine solution using electrochemical impedance spectroscopy, selected growth defects were analyzed by the SEM/FIB. Using FIB, we made serial cross-sections through individual selected defects in order to obtain an insight into the corrosion process at different kinds of growth defects.

We confirmed that pitting corrosion is closely related to the through-thickness growth defects. We found that the intensity of the pitting corrosion of different through-thickness growth defects is very different. On the same sample, exposed to corrosive medium, the effects of pitting corrosion at the sites of some defects are hardly visible, while at others, catastrophic corrosion attacks occurred. It all depends on how very open the path to the substrate is. We also found that in the case of nodular defects, the intensity of corrosion depends on the shape of the seed. The shape of the seed determines whether the crevice down to the substrate will occur or not.

Author Contributions: Conceptualization, Manuscript Writing, P.P.; FIB Analysis of Growth Defects and Corroded Samples, A.D.; SEM Analysis of Defects, P.G.; Manuscript Review and Editing, M.Č. and M.P.; FIB analysis of Nonmetallic Inclusions, T.B.; Electrochemical Impedance Test, D.K.M.

Funding: This work was funded by the Slovenian Research Agency (program P2-0082). We also acknowledge funding from the European Regional Development Funds (CENN Nanocenter, OP13.1.1.2.02.006).

Acknowledgments: The authors wish to commemorate Darja Kek Merl who passed away two years ago.

Conflicts of Interest: The authors declare no conflict of interest.

References

1. Tietems, R. Large-scale industrial coating applications and systems comprehensive materials processing. *Elsevier* **2014**, *4*, 519–561.
2. Haubner, R.; Lessiak, M.; Pitonak, R.; Köpf, A. Weissenbacher. *Int. J. Refract. Met. Hard Mater.* **2017**, *62*, 210–218. [CrossRef]
3. D'Avico, L.; Beltrami, R.; Lecis, N.; Trasatti, S.P. Corrosion behavior and surface properties of PVD coatings for mold technology applications. *Coatings* **2019**, *9*, 7. [CrossRef]
4. Jehn, H.A. Improvement of the corrosion resistance of PVD hard coating–substrate systems. *Surf. Coat. Technol.* **2000**, *125*, 212–217. [CrossRef]
5. Fenker, M.; Balzer, M.; Kappl, H. Corrosion protection with hard coatings on steel: Past approaches and current research efforts. *Surf. Coat. Technol.* **2014**, *257*, 182–205. [CrossRef]
6. Ahn, S.; Lee, J.; Kim, J.; Han, J. Localized corrosion mechanisms of the multilayered coatings related to growth defects. *Surf. Coat. Technol.* **2004**, *177*, 638–644. [CrossRef]
7. Balzer, M. Identification of the growth defects responsible for pitting corrosion on sputter-coated steel samples by Large Area High Resolution mapping. *Thin Solid Films* **2015**, *581*, 99–106. [CrossRef]
8. Lewis, D.B.; Creasey, S.J.; Wüstefeld, C.; Ehiasarian, A.P.; Hovsepian, P.E. The role of the growth defects on the corrosion resistance of CrN/NbN superlattice coatings deposited at low temperatures. *Thin Solid Films* **2006**, *503*, 143–148. [CrossRef]
9. Korhonen, A.S. Corrosion of thin hard PVD coatings. *Vacuum* **1994**, *45*, 1031–1034. [CrossRef]
10. Merl, D.K.; Panjan, P.; Panjan, M.; Čekada, M. The role of surface defects density on corrosion resistance of pvd hard coatings. *Plasma Process. Polym.* **2007**, *4*, 5613–5617. [CrossRef]

11. Penttinen, I.M.; Korhonen, A.S.; Harju, E.; Turkia, M.A.; Forsen, O.; Ristolainen, O.E. Comparison of the corrosion resistance of TiN and (Ti,Al)N coatings. *Surf. Coat. Technol.* **1992**, *50*, 161–168. [CrossRef]
12. Wang, H.W.; Stack, M.M.; Lyon, S.B.; Hovsepian, P.; Münz, W.-D. The corrosion behaviour of macroparticle defects in arc bond-sputtered CrN/NbN superlattice coatings. *Surf. Coat. Technol.* **2000**, *126*, 279–287. [CrossRef]
13. Hoche, H.; Pusch, C.; Oechsner, M. Establishing PVD-coatings for the corrosion protection of mild steel substrates for complex tribological and corrosive stresses. *Surf. Coat. Technol.* **2018**, in press. [CrossRef]
14. Mattox, D.M. Surface effects on the growth, adhesion and properties of reactively deposited hard coatings. *Surf. Coat. Technol.* **1996**, *81*, 8–16. [CrossRef]
15. Liu, C.; Leyland, A.; Bi, Q.; Matthews, A. Corrosion resistance of multilayered plasma-assisted physical vapour deposition TiN and CrN coatings. *Surf. Coat. Technol.* **2001**, *141*, 164–173.
16. Fenker, M.; Balzer, M.; Kappl, H. Corrosion behaviour of decorative and wear resistant coatings on steel deposited by reactive magnetron sputtering–Tests and improvements. *Thin Solid Films* **2006**, *515*, 27–32. [CrossRef]
17. Härkönen, E.; Kolev, J.; Díaz, B.; Swiatowska, J.; Maurice, V.; Seyeux, A.; Marcus, P.; Fenker, M.; Toth, L.; Radnoczi, G.; et al. Sealing of hard CrN and DLC coatings with atomic layer deposition. *ACS Appl. Mater. Interfaces* **2014**, *6*, 1893–1901. [CrossRef]
18. Wan, Z.; Zhang, T.F.; Ding, J.C.; Kim, C.-M.; Park, S.-W.; Yang, Y.; Kim, K.-H.; Kwon, S.-H. Enhanced corrosion resistance of pvd-crn coatings by ald sealing layers. *Nanoscale Res. Lett.* **2017**, *12*, 248. [CrossRef]
19. Heyn, A.; Mueller, T.; Balzer, M.; Kappl, H.; Fenkel, M. Corrosion protection mechanisms of TiMgN hard coatings on steel. *IOP Conf. Ser. Mater. Sci. Eng.* **2018**, *373*, 012009. [CrossRef]
20. Barshilia, H.C.; Prakash, M.S.; Poojari, A.; Rajam, K.S. Corrosion behavior of nanolayered TiN/NbN multilayer coatings prepared by reactive direct current magnetron sputtering process. *Thin Solid Films* **2004**, *460*, 133–142. [CrossRef]
21. Herranen, M.; Wiklund, U.; Carlsson, J.-O.; Hogmark, S. Corrosion behaviour of Ti/TiN multilayer coated tool steel. *Surf. Coat. Technol.* **1998**, *99*, 191–196. [CrossRef]
22. Mendibide, C.; Steyer, P.; Millet, J.-P. Formation of a semiconductive surface film on nanomultilayered TiN/CrN coatings and its correlation with corrosion protection of steel. *Surf. Coat. Technol.* **2005**, *200*, 109–112. [CrossRef]
23. Abdeen, D.H.; el Hachach, M.; Koc, M.; Atieh, M.A. A review on the corrosion behaviour of nanocoatings on metallic substrates. *Materials* **2019**, *12*, 210. [CrossRef]
24. Fenker, M.; Balzer, M.; Jehn, H.A.; Kappl, H.; Lee, J.J.; Lee, K.H.; Park, H.S. Improvement of the corrosion resistance of hard wear resistant coatings by intermediate plasma etching or multilayered structure. *Surf. Coat. Technol.* **2002**, *150*, 101–106. [CrossRef]
25. Abusuilik, S.B.; Inoue, K. Effects of intermediate surface treatments on corrosion resistance of cathodic arc PVD hard coatings. *Surf. Coat. Technol.* **2013**, *237*, 421–428. [CrossRef]
26. Fenker, M.; Kappl, H.; Petrikowski, K.; Bretzler, R. Pulsed power magnetron sputtering of a niobium target in reactive oxygen and/or nitrogen atmosphere. *Surf. Coat. Technol.* **2005**, *200*, 1356. [CrossRef]
27. Reinhard, C.; Ehiasarian, A.P.; Hovsepian, P.E. CrN/NbN superlattice structured coatings with enhanced corrosion resistance achieved by high power impulse magnetron sputtering interface pre-treatment. *Thin Solid Films* **2007**, *515*, 3685–3692. [CrossRef]
28. Panjan, P.; Čekada, M.; Panjan, M.; Kek-Merl, D. Growth defects in PVD hard coatings. *Vacuum* **2010**, *84*, 209–214. [CrossRef]
29. Panjan, P.; Merl, D.K.; Zupanič, F.; Čekada, M.; Panjan, M. SEM study of defects in PVD hard coatings using focused ion beam milling. *Surf. Coat. Technol.* **2008**, *202*, 2302–2305. [CrossRef]
30. Harlin, P.; Bexell, U.; Olsson, M. Influence of surface topography of arc-deposited TiN and sputter-deposited WC/C coatings on the initial material transfer tendency and friction characteristics under dry sliding contact conditions. *Surf. Coat. Technol.* **2009**, *203*, 1748–1755. [CrossRef]
31. Balzer, M.; Fenker, M.; Kappl, H.; Muller, T.; Heyn, A.; Heiss, A.; Richter, A. Corrosion protection of steel substrates by magnetron sputtered TiMgN hard coatings: Structure, mechanical properties and growth defect related salt spray test results. *Surf. Coat. Technol.* **2018**, *349*, 82–92. [CrossRef]

32. Merl, D.K.; Panjan, P.; Čekada, M.; Maček, M. The corrosion behavior of Cr-(C,N) PVD hard coatings deposited on various substrates. *Electrochim. Acta* **2004**, *49*, 1527–1533. [CrossRef]

coatings

MDPI

Article

Short- and Long-Term Wettability Evolution and Corrosion Resistance of Uncoated and Polymer-Coated Laser-Textured Steel Surface

Marjetka Conradi [1,*], Tina Sever [1], Peter Gregorčič [2] and Aleksandra Kocijan [1]

[1] Institute of metals and technology, Lepi pot 11, 1000 Ljubljana, Slovenia; tina.sever@imt.si (T.S.); Aleksandra.kocijan@imt.si (A.K.)

[2] Faculty of Mechanical Engineering, University of Ljubljana, Aškerčeva 6, 1000 Ljubljana, Slovenia; Peter.Gregorcic@fs.uni-lj.si

* Correspondence: marjetka.conradi@imt.si; Tel.: +386-1-4701-955; Fax: +386-1-4701-927

Received: 28 August 2019; Accepted: 17 September 2019; Published: 19 September 2019

Abstract: We present the results of one year observation of wetting and corrosion properties of nanosecond fiber laser-textured stainless steel, uncoated and coated with epoxy or FAS (fluoroalkylsilane)-TiO_2/epoxy. A comparative study was performed on samples kept under ambient conditions and in reduced air pressure and humidity. The results show the ability to induce wettability conversion from initially superhydrophilic to final superhydrophobic state either indirectly by ageing the uncoated laser-textured surface or directly by application of FAS-TiO_2/epoxy coating. The storage conditions significantly influenced the wettability development of uncoated laser-textured steel, i.e., the process of ageing was slowed down in reduced air pressure and humidity. Detailed surface chemical analysis revealed that adsorption of the organic matters from the surrounding media influences the wettability conversion and ageing. However, the ageing of the coated surfaces was not affected by the storage conditions. Corrosion stability of uncoated laser-textured surfaces was enhanced over time due to the wettability transition, depending on their morphology. Coatings represent a superior barrier over the texture and wettability with the stable long-term surface protection against aggressive media.

Keywords: stainless steel; laser surface engineering; wettability conversion; surface chemistry; polymer coatings

1. Introduction

Austenitic stainless steel is known as one of the most important engineering materials due to its high corrosion resistance combined with favorable mechanical properties, such as high tensile strength [1,2]. It is widely used in several applications in medicine, the food industry, and environmental or construction engineering. However, for applications in aquatic media or in other aggressive environments, such as a chloride ion-rich environment, it is extremely important to adjust the wetting properties of the steel surface to ensure its optimal performance. This prolongs the lifetime of the material, prevents corrosion, and consequently reduces the additional costs if the material cannot fulfil the demands of the operating conditions. Manipulation of surface wettability of metallic surfaces, therefore, still represents a scientific challenge in view of optimization of surface properties for specific end uses within economically acceptable limitations.

Generally, metals are covered with an oxide layer that tends to be (super)hydrophilic due to high surface energy of metallic oxides [3]. Therefore, to turn a surface (super)hydrophobic, the surface energy of metallic oxide surface has to be lowered. In this respect, one of the most addressed procedures is surface chemical modification by surface coatings [4,5]. Research has been focused mostly on synthesis of multifunctional superhydrophobic coatings for water-repellent and self-cleaning applications [6–8], where (super)hydrophobicity is regulated through multiscale roughness structures formed by various nanoparticles, i.e., silica, TiO_2, ZnO, etc. [9–11]. Polymer coatings, on the other hand, present an effective physical barrier between the metal and the environment containing aggressive species, and serve as a reservoir for corrosion inhibitors that help the steel surface to resist attacks from aggressive species [12].

Recently, laser texturing has been addressed as one of the most promising approaches, in view of its flexibility, accuracy, chemical-free production, lack of tool wear, and negligible effect on the properties of bulk material. Several studies report on micro-, nano- or picosecond laser-structured metallic surfaces which are initially (super)hydrophilic and turn (super)hydrophobic in ambient conditions with time due to the adsorption of organic matters from the surrounding atmosphere over time [3,5,13–19].

The wettability conversion is strongly correlated to enhancing corrosion protective performance of surfaces [20,21]. It has been shown that the most effective and durable protection against corrosion in aggressive media is provided by superhydrophobic surfaces/coatings in combination with pre-existing oxide layers typical for specific metallic substrates [20]. Although several studies have been devoted to the examination of how processing parameters influence metallic-surface wettability and corrosion behavior, most of them were performed only within a short-term (two month) period [17,18,21,22]. In our previous work [19], we have shown that solely the short-term evaluation of superhydrophilic-to-hydrophobic transition after the laser texturing of metallic surface may lead to wrong conclusions. However, there is still a lack of analysis about how the transition of wettability influences the electrochemical properties and corrosion stability of uncoated or coated laser textured surfaces within the long-term (one year) period.

In this paper, we studied short- and long-term wettability evolution and corrosion resistance of uncoated laser-textured surfaces and laser-textured surfaces coated with epoxy and FAS-TiO_2/epoxy composite coating. We compared the ageing of surface properties of samples stored in two atmospheres: in ambient conditions and in a chamber with reduced air pressure and humidity. By monitoring the static water contact angles and detailed surface chemical analysis we evaluated the time-dependence of the wettability transition from a superhydrophilic to superhydrophobic state of uncoated laser-textured surfaces, as well as hydrophobic to superhydrophobic transition of coated laser-textured surfaces. Short- and long-term protective performance of uncoated and coated surfaces in chloride ion-rich environments were analyzed with potentiodynamic measurements.

2. Materials and Methods

Materials—Austenitic stainless steel AISI 316L (17% Cr, 10% Ni, 2.1% Mo, 1.4% Mn, 0.38% Si, 0.041% P, 0.021% C, <0.005% S in mass fraction) was used as a substrate. The steel sheet with a thickness of 1.5 mm was cut into discs of 25 mm diameter, which were cleaned with ethanol in ultrasonic bath.

A two-component USP Class VI biocompatible epoxy EPO-TEK 302-3M (EPOXY TECHNOLOGY, Inc., Billerica, MA, USA) was mixed in the wt.% ratio 100:45 and used as a coating. TiO_2 nanoparticles with mean diameters of 30 nm were provided by Degussa, Wilmington USA. Prior to coating preparation, nanoparticles were functionalized in 1 vol % ethanolic fluoroalkylsilane (Sigma-Aldrich, Darmstadt, Germany) or FAS ($C_{16}H_{19}F_{17}O_3Si$) solution to induce hydrophobic effect.

Laser Texturing of the Surfaces—The surfaces were laser-textured by using a fiber laser (SPI Lasers, Ltd., G4, SP-020P-A-HS-S-A-Y, Southampton, UK) with a wavelength of 1060 nm, pulse duration of 28 ns, and pulse energy of 0.2 mJ. The laser beam was led through a scanning head (Raylase, SS-IIE-10, Wessling, Germany) and an F-theta lens by a focal distance of 163 mm. Since the beam expander to attain a beam diameter on the lens of 7.5 mm was used, the beam spot size on the sample, placed in a focal position, equals 38 m (considering the beam quality parameter $M^2 = 1.3$). The beam was led over the surface, first in y (90°) and then in x direction (0°), by scanning velocity of 270 mm/s. The pulse repetition rate equaled 90 kHz. Thus, the two consecutive pulses were separated by 3 m

(92% overlapping). The scanning line separations were equal for x and y direction. The following different scanning line separations were used: $x = y = 25, 50, 100$ m.

Coatings Preparation—Two types of coatings were spin-coated on laser-textured steel surfaces: epoxy coating (E) and FAS-TiO_2 nanoparticle epoxy coating (FAS-TiO_2/E). FAS-TiO_2/E coatings were prepared by first spin coating 2 wt % ethanolic solution of FAS-TiO_2 nanoparticles, which were covered with epoxy layer. All the coatings were cured in an oven for 3 h at 65 °C.

Surface Characterization—Scanning electron microscopy (SEM) analysis using field emission SEM JEOL JSM-6500F (JEOL, Tokyo, Japan) was employed to investigate the morphology of the laser-textured surfaces and the laser-textured surfaces coated with E and FAS-TiO_2/E coating.

The X-ray photoelectron spectroscopy (XPS) analyses were carried out on the PHI-TFA XPS spectrometer produced by Physical Electronics Inc (Chanhassen, MN, USA). The analyzed area was 0.4 mm in diameter and the analyzed depth was about 3–5 nm. Sample surfaces were excited by X-ray radiation from monochromatic Al source at photon energy of 1486.6 eV. The survey wide energy spectra were taken with pass energy of analyzer of 187 eV in order to identify and quantify present elements on the surface. The high-energy resolution spectra were acquired with energy analyzer operating at resolution of about 0.6 eV and pass energy of 29 eV. Casa XPS® software (version 2.3.15) [23] was used for the detailed data processing.

The surface wettability was evaluated with static water contact angles (θ) on laser-textured, epoxy-coated, and FAS-TiO_2/epoxy-coated laser-textured surfaces by using a surface energy evaluation system (Advex Instruments s.r.o., Brno, Czech Republic). Water drops of 5 μL were deposited on at least three different spots of the substrates to avoid the influence of roughness and gravity on the shape of the drop. The drop contour was analyzed from the image of the deposited liquid drop on the surface and the contact angle was determined by using Young–Laplace fitting. To minimize the errors due to roughness and heterogeneity, the average values of the contact angles of the drop were measured approximately 20 s after the deposition. All the contact angle measurements were carried out at 22 °C and ambient humidity.

Optical 3D metrology system, model Alicona Infinite Focus (Alicona Imaging GmbH, Graz, Austria), was used for the laser-textured surface topography analysis. The 3D surface roughness parameters, including the average height of the selected area, S_a, were evaluated and calculated from the Alicona Infinite Focus using IF-MeasureSuite Version 5.1. At least three measurements per sample were performed at magnification 200×, with a lateral resolution of 0.9 μm and a vertical resolution of 50 nm. The size of the analyzed area was 1335 × 997 μm^2.

Electrochemical Measurements—Electrochemical measurements were carried out in a simulated physiological Hank's solution, containing 8 g/L NaCl, 0.40 g/L KCl, 0.35 g/L $NaHCO_3$, 0.25 g/L $NaH_2PO_4 \times 2H_2O$, 0.06 g/L $Na_2HPO_4 \times 2H_2O$, 0.19 g/L $CaCl_2 \times 2H_2O$, 0.41 g/L $MgCl_2 \times 6H_2O$, 0.06 g/L $MgSO_4 \times 7H_2O$, and 1 g/L glucose, at pH = 7.8 and room temperature. The solution was chosen for possible biomedical applications. We used three-electrode flat corrosion cells. The investigated specimen was employed as working electrode (WE), a saturated calomel electrode (SCE, 0.242 V vs. SHE, standard hydrogen electrode) as reference electrode (RE), and a platinum net as counter electrode (CE). The measurements were performed by using BioLogic Modular Research Grade Potentiostat/Galvanostat/FRA Model SP-300 with an EC-Lab Software (version 11.27). The potentiodynamic curves were recorded after 1 h sample stabilization at the open-circuit potential (OCP), starting the measurement at 250 mV vs. SCE more negative than the OCP. The potential was then increased, using a scan rate of 1 mV s^{-1}. All the measurements were repeated three times.

3. Results and Discussion

3.1. Surface Morphology

The modification of as-received stainless steel surface morphology after laser-texturing with different scanning line separations Δx and Δy (0°/90°) is shown in Figure 1. The 0°/90° scanning strategy was used to obtain the same morphology in x and y direction and to ensure that the results are comparable to our previous work [21]. It is clearly visible that for scanning line separation $\Delta x = \Delta y = 100$ µm well-defined µ-channels with approximate size of the directly laser ablated area of 40-µm width were formed (Figure 1 a,b). For $\Delta x = \Delta y = 50$ µm, the µ-channels start to overlap but we can still recognize the x-y µ-channel pattern (Figure 1c,d). As already shown [19], the samples with the smallest scanning line separation, $\Delta x = \Delta y = 25$ µm, result in a rough surface with no characteristic µ-channel pattern (Figure 1e,f). A similar effect was observed also in the case of 0° scanning strategy with small scanning line separations [24].

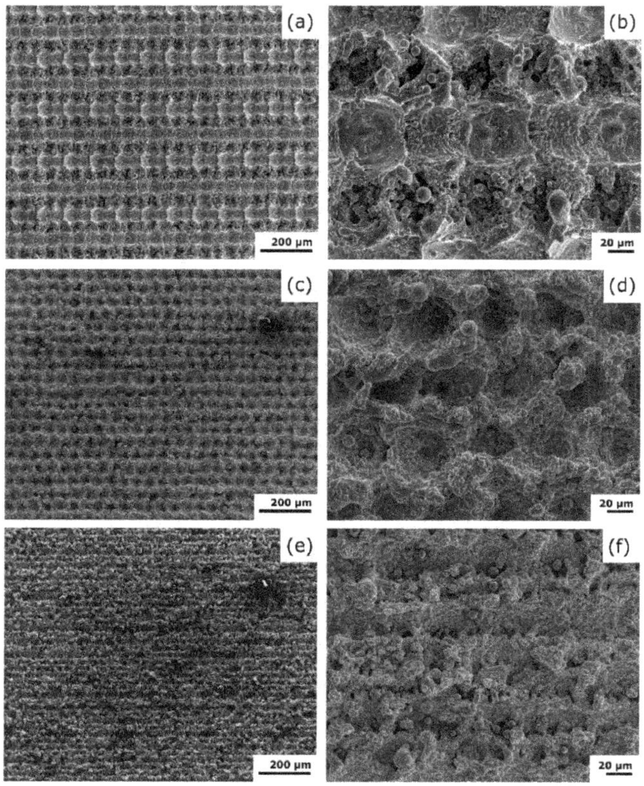

Figure 1. SEM micrographs of the laser-textured surfaces with different scanning line separations: (**a**,**b**) $\Delta x = \Delta y = 100$ µm, (**c**,**d**) $\Delta x = \Delta y = 50$ µm, (**e**,**f**) $\Delta x = \Delta y = 25$ µm.

After laser-texturing, some of the surfaces were additionally coated with pure epoxy coating (E) and some with epoxy coating enriched with FAS-functionalized 30 nm TiO_2 nanoparticles (FAS-TiO_2/E).

The average surface roughness parameter, S_a, and the rough surface areas were measured with Alicona Imaging System to evaluate the topographical differences between the laser-textured steel surfaces. As revealed in Table 1, the average surface roughness for the surfaces with $\Delta x = \Delta y = 100$ and 50 µm is of the same order of magnitude, $S_a = 10.8 \pm 0.1$ µm. S_a, however, drops for a factor of two for the surface with the smallest scanning line separation $\Delta x = \Delta y = 25$ µm. This indicates the increase of the laser-textured surface porosity, which is a consequence of the overlapped µ-channels [24]. The average surface roughness is, on the other hand, not affected by the application of epoxy and FAS-TiO_2/E coatings.

Table 1. Average surface roughness parameter S_a and the ratio between the rough and the smooth surface area, A_R/A_0 of uncoated and coated laser-textured surfaces.

Scan Line Separation $\Delta x = \Delta y$ [µm]	S_a [µm]	A_R/A_0
100 µm	10.8 ± 0.1	2.0 ± 0.3
100 µm, E	10.9 ± 0.1	1.7 ± 0.4
100 µm, FAS-TiO_2/E	11.2 ± 0.1	1.8 ± 0.2
50 µm	10.7 ± 0.1	4.3 ± 0.1
50 µm, E	10.4 ± 0.2	4.3 ± 0.2
50 µm, FAS-TiO_2/E	10.7 ± 0.2	4.1 ± 0.2
25 µm	5.2 ± 0.1	1.5 ± 0.1
25 µm, E	5.3 ± 0.2	1.4 ± 0.2
25 µm, FAS-TiO_2/E	5.6 ± 0.2	1.6 ± 0.2

The rough surface area was estimated through the ratio A_R/A_0 between the area of the rough, laser-textured surface, A_R and the smooth (unprocessed) surface, A_0. For the polished surface, the ratio $A_R/A_0 = 1$ and is increased due to laser ablation after laser texturing. This increase is important in characterization of the correlation between the developed rough surface and corrosion. The largest surface area results in a surface processed with $\Delta x = \Delta y = 50$ µm, while the surfaces processed with $\Delta x = \Delta y = 100$ and 25 µm have two to three times smaller surface, compared to $\Delta x = \Delta y = 50$ µm surface (Table 1).

3.2. *Surface Wettability and Aging*

We evaluated short-term (within three months) and long-term (after one year) development of the static water contact angles on uncoated, epoxy-coated and FAS-TiO_2/E-coated surfaces processed with scanning line separation $x = y = 100$, 50, and 25 m. The wettability evolution was compared for samples stored in two different media, in ambient conditions and in a chamber with reduced air pressure and humidity.

3.2.1. Uncoated Laser-Textured Surfaces

Similarly as in Ref. [15], the static water contact angle measurement immediately after the laser texturing showed that the surfaces were superhydrophilic in a saturated Wenzel regime ($\theta = 0°$) [18,25] for all scanning line separations. This means that the water droplets spilled into a thin film and completely wet the laser-textured surface. As shown in Figure 2, after initial superhydrophilicity, the surface wettability eventually starts to decrease, which is reflected in the increase of the contact angle. The surface first turns hydrophilic, then hydrophobic, until long-term stability of the superhydrophobicity is achieved for all scanning line separations in both storage media. This suggests a transition from Wenzel to Cassie–Baxter state due to the incomplete surface wetting after the static water contact angles saturate in a steady state [17]. Each measurement point in the graph represents the average of at least three individual measurements.

The contact angle evolution for all the samples can be well-fitted with exponential growth function, as shown in Figure 2, similar to Ref. [18]: $\frac{\theta}{\theta_0} = 1 - e^{-t/\tau}$. Here, θ_0 and τ are steady state contact angle and time constant, respectively, and are listed in Figure 2 and Table 2. The points were fitted from the day when the contact angle was above 0°.

Even though we observed a similar trend in wettability development for all the surfaces, there was a clear difference in the progress of (super)hydrophobicity, depending upon scanning line separation as well as different storage media. The surface processed with $\Delta x = \Delta y = 100$ µm and stored in ambient conditions started to turn hydrophobic (measured non-zero contact angle) two days after laser texturing. Afterwards, we observed a characteristic 16-day transition period of hydrophobicity development with increasing contact angles of typically irregularly shaped droplets. An example of such irregularly shaped droplet development is shown in Figure 3; however it will be discussed in detail in an independent study. Finally, the surface turned hydrophobic with observed spherically shaped droplets until reaching an equilibrium steady state superhydrophobicity.

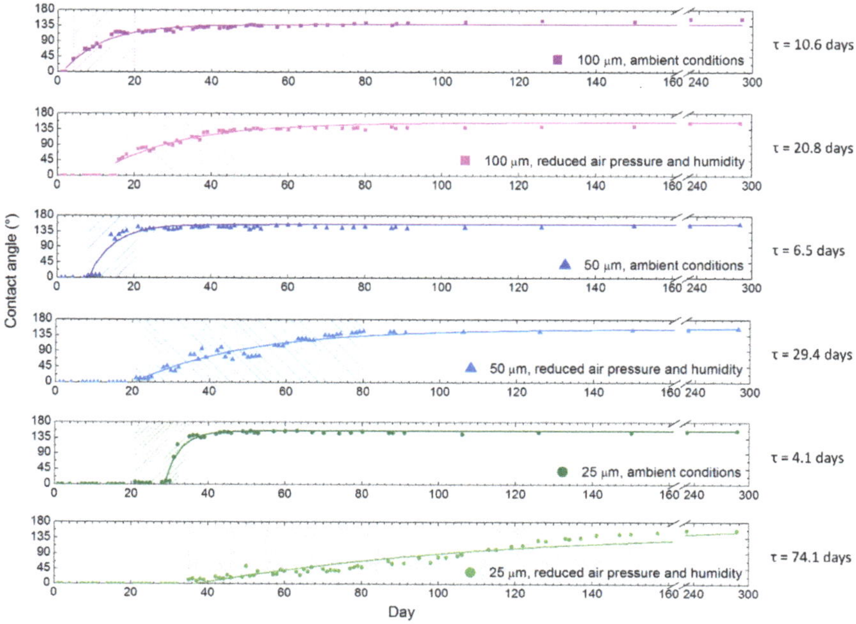

Figure 2. Short- and long-term development of static water contact angles with characteristic exponential growth fits and corresponding time constants of uncoated laser-textured steel surfaces with different scanning line separations ($\Delta x = \Delta y = 100, 50, 25$ μm), stored in ambient conditions and in a chamber with reduced air pressure and humidity.

Table 2. Comparison of wettability development with corresponding time constants on samples stored in ambient conditions and in a chamber with reduced air pressure and humidity: time evaluation of complete wetting regime and transition period.

Scan Line Separations [μm]	Complete Wetting Regime	Transition Period	Time Constant, τ
100, ambient	4 days	16 days	10.6 days
100, chamber	15 days	31 days	20.8 days
50, ambient	8 days	13 days	6.5 days
50, chamber	21 days	59 days	29.4 days
25, ambient	21 days	14 days	4.1 days
25, chamber	35 days	122 days	74.1 days

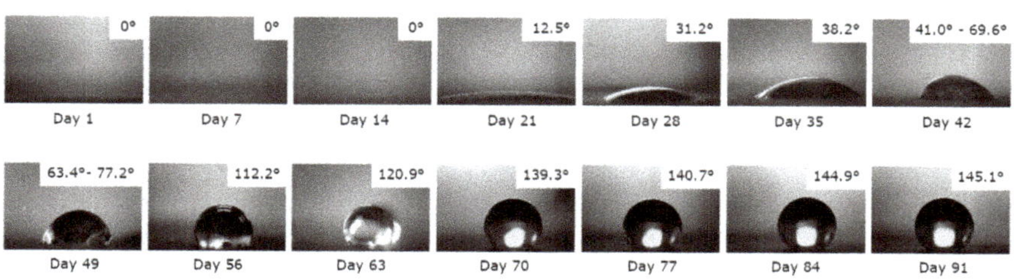

Figure 3. A typical example of droplet evolution from superhydrophilic to superhydrophobic state, through a phase with characteristic irregularly shaped droplets, on uncoated laser-textured surface with $\Delta x = \Delta y = 50$ μm.

Surfaces stored in ambient conditions and processed with $\Delta x = \Delta y = 50$ and 25 μm started to turn hydrophobic after 8 and 21 days, respectively. On both samples, hydrophobicity was developing for approximately two weeks, similarly as described above, until the contact angle gradually reached a steady state contact angle of 146.3° ± 2.1° on a surface with $\Delta x = \Delta y = 50$ μm, whereas surface with $\Delta x = \Delta y = 25$ μm became superhydrophobic with a contact angle of 150.7° ± 2.3°. The two week transition period is also for surfaces with $\Delta x = \Delta y = 50$ and 25 μm, characterized with irregularly shaped droplets (Figure 3).

In a chamber with reduced air pressure and humidity, we observed that the process of hydrophobicity development of uncoated laser-textured surfaces for all scanning line separations was significantly slowed down, which is reflected also in larger time constants. As shown in Figure 2 and Table 2, hydrophobicity starts to develop with, approximately, a two week delay for all the samples stored in a chamber compared to the corresponding samples stored in ambient conditions. Afterwards, the hydrophobicity on samples stored in a chamber starts to increase, similarly as on samples stored in the ambient conditions. However, the time needed to achieve the saturated contact angle (the transition period) was prolonged (Table 2).

These results suggest that scanning line separation, together with the reduced air pressure and humidity, slow down surface ageing and the wettability conversion. In ambient conditions, the decrease of the scanning line separation postponed the start of the hydrophobicity development. However, the transition period before reaching saturation of the contact angle values was approximately the same for all the samples, i.e., 2 weeks (Table 2). Besides that, the smaller the scanning line separation, the steeper the increase of the contact angle to the steady state. In a chamber, the start of the hydrophobicity development is postponed with the decrease of the scanning line separation in the same manner as in ambient conditions.

To understand the mechanism of the ageing of uncoated laser-textured steel surfaces with time-dependent wettability conversion from superhydrophilic to superhydrophobic state, comparison of XPS surface chemical analysis of samples immediately after the preparation and after one year was evaluated. We performed deconvolution of C 1*s*, O 1*s*, Fe $2p_{3/2}$, Cr $2p_{3/2}$, Mo 3*d*, and Ni 2*p* peaks to analyze the surface state and the amount of carbon and other elements in the top layer. Deconvolution of all C peaks was performed by using fitting procedure after subtraction of a Shirley's background correction. The shape of all peaks was kept to be Gaussian/Lorentzian 60/30 ratio. C 1*s* peak was fitted with three species. The main peak around 284.7 eV was attributed to the hydrocarbon chains (C–C/C–H). The C–O represented the carbon bonded to one oxygen and C=O represented the double bond between carbon and oxygen, at 1.4–1.5 and 2.8–3.1 eV from the C–C/C–H components, respectively [26,27]. Figure 4 presents the examples of the deconvoluted XPS spectra of C 1*s*, O 1*s*, Fe $2p_{3/2}$, and Cr $2p_{3/2}$ laser-textured AISI 316L surface with $\Delta x = \Delta y = 50$ μm stored in ambient conditions immediately after preparation.

As shown in Figure 5, the composition of surface functional groups of laser-textured surfaces ($\Delta x = \Delta y = 25$, 50, and 100 μm) changed considerably on aged samples after one year, compared to samples immediately after preparation. To see these changes, the relative amounts of nonpolar C–C/C–H groups and metal oxides (M–O) were evaluated. The results show significantly higher amount of C–C/C–H groups on aged samples (Figure 5a). Due to their nonpolar nature, they serve as hydrophobic regions on the surface and therefore influence the wettability greatly. We should however stress that in spite of the presence of minor polar peak attributed to C–O and C=O groups on laser-textured surface immediately after the preparation, as well as after one year (Figure 5c), which serve as hydrophilic regions, wettability conversion from the initial superhydrophilicity to the final superhydrophobicity was observed on all uncoated laser-textured samples. Initial superhydrophilicity of laser-textured steel surfaces can be explained by a large number of polar sites originating from the presence of surface metal oxides, the amount of which decreases after one year (Figure 5b) [5,18]. The detected amount of nonpolar carbon immediately after the preparation is too small to shield the hydrophilicity of the surface oxides. Therefore, the intrinsic hydrophilicity of the polar metal oxides is amplified by the laser-induced surface roughness according to Wenzel's theory. The conversion of surface wettability, on the other hand, is known to be caused by the adsorption of organic matters from the surrounding atmosphere over time [5,28]. This was confirmed by evaluating the atomic ratio between carbon and metal species (C/M), which is governed by the amount of adsorbed organic matters (Figure 5b). Finally, higher amount of C–C/C–H groups, lower amount of metal oxides, and higher C/M ratio on aged samples lead to more nonpolar (super)hydrophobic surface (Figure 5a,d).

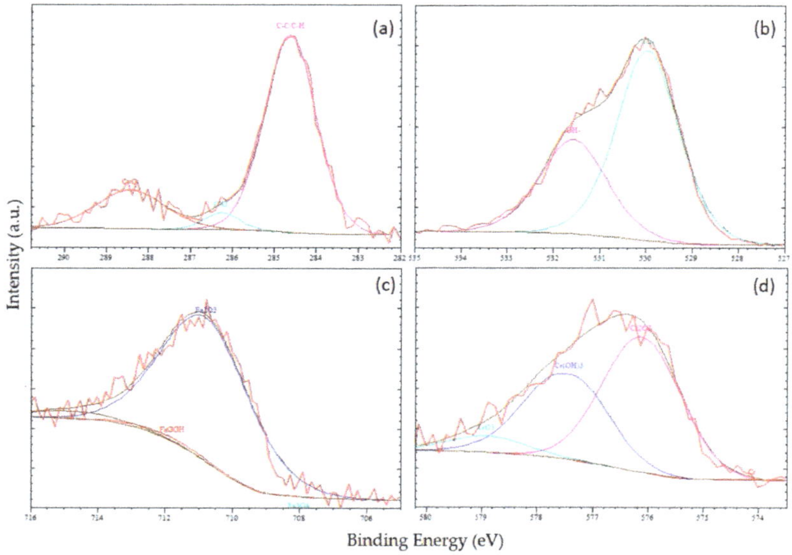

Figure 4. Deconvoluted XPS spectra of C 1s (**a**), O 1s (**b**), Fe $2p_{3/2}$ (**c**), Cr $2p_{3/2}$ (**d**) laser-textured AISI 316L surface $\Delta x = \Delta y = 50$ μm stored in ambient conditions immediately after preparation.

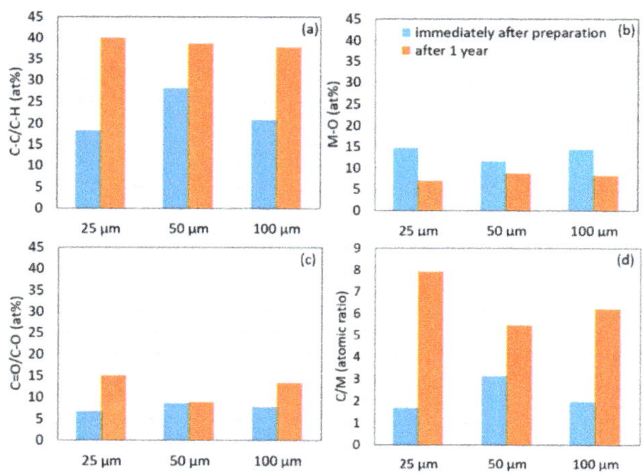

Figure 5. Relative amount of C–C/C–H functional groups (at.%) (**a**), relative amount of metal oxides M–O (**b**), relative amount of C=O/C–O (**c**), and the ratio of C/M obtained from the high-resolution XPS spectra (**d**) for laser-textured surfaces with $\Delta x = \Delta y = 25$, 50, and 100 μm.

3.2.2. Coated Laser Textured Surfaces

On coated laser-textured steel surfaces, no wettability conversion was observed. Depending on the scan line separation, the laser-textured surfaces coated with epoxy were immediately after preparation poorly (for $\Delta x = \Delta y = 100$ µm, 97.5° ± 1.4°) to moderately hydrophobic (for $\Delta x = \Delta y = 50$ and 25 µm; 112.1° ± 1.9° and 108.3° ± 2.3°, respectively). FAS-TiO_2/E-coated laser-textured surfaces were, however, strongly hydrophobic for $\Delta x = \Delta y = 100$ µm (143.4° ± 3.1°) and superhydrophobic for $\Delta x = \Delta y = 50$ and 25 µm (151.1° ± 2.1° and 151.9° ± 0.7°, respectively).

Short- and long-term contact angle development on epoxy-coated and FAS-TiO_2/E-coated laser-textured steel surfaces in ambient conditions and in a chamber with reduced air pressure and humidity is presented in Figure 6. Epoxy-coated samples stored in air and in a chamber remained poorly hydrophobic with a measured slight increase (few degrees, <10°) of contact angles on a short- and long-term basis for all scanning line separations (Table 3). FAS-TiO_2/E-coated surfaces in both media and for all scanning line separations turned or remained superhydrophobic on a short- and long-term basis again with a slight increase of contact angles (<5°) after a one year measurement period (Table 3). These results, therefore, indicate the long-term stability of coatings with stable and predictable wettability.

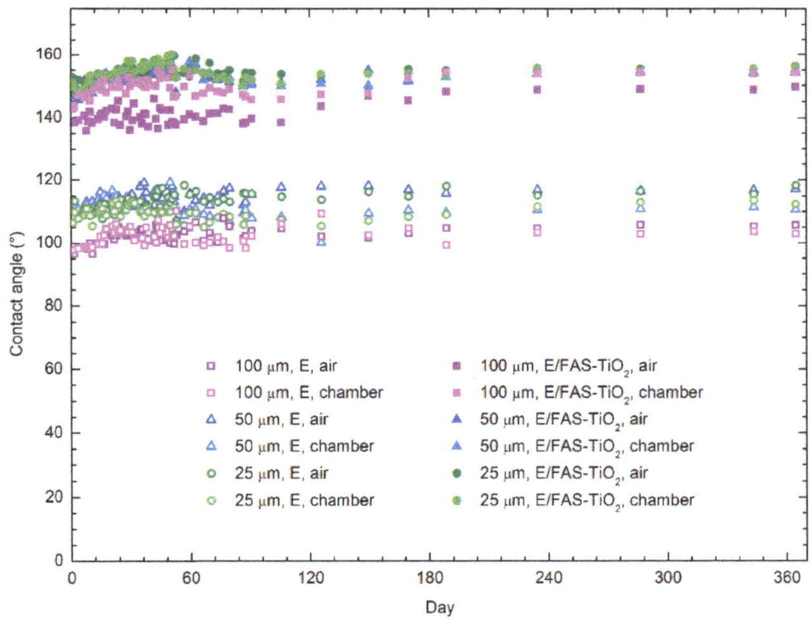

Figure 6. Time development of static water contact angles of epoxy-coated and FAS-TiO_2/E-coated laser-textured steel surfaces for different scan line separations ($\Delta x = \Delta y = 100$, 50, 25 µm) stored in ambient conditions and in a chamber with reduced air pressure and humidity.

Table 3. Comparison of static water contact angles of uncoated, epoxy-, and FAS-TiO_2/E-coated samples with different scan line separations ($\Delta x = \Delta y = 100, 50, 25$ µm) measured immediately after preparation, after 3 months, and after 1 year, when stored in ambient conditions and in a chamber with reduced air pressure and humidity.

Scan Line Separation [µm]	Uncoated Contact angle [°]	E Contact Angle [°]	FAS-TiO_2/E Contact Angle [°]
Immediately after preparation			
100, ambient	0	97.5 ± 1.4	143.4 ± 3.1
50, ambient	0	112.1 ± 1.9	151.1 ± 2.1
25, ambient	0	108.3 ± 2.3	151.9 ± 0.7
3 months after preparation			
100, ambient	130.5 ± 0.6	104.3 ± 2.6	140.5 ± 2.2
100, chamber	138.6 ± 3.2	100.2 ± 1.7	147.3 ± 1.2
50, ambient	146.3 ± 1.6	115.0 ± 2.2	151.1 ± 0.9
50, chamber	145.1 ± 1.6	108.8 ± 1.5	151.5 ± 1.0
25, ambient	150.7 ± 1.1	115.0 ± 1.2	153.7 ± 1.3
25, chamber	56.5 ± 7.8	107.3 ± 2.2	152.0 ± 1.1
1 year after preparation			
100, ambient	154.5 ± 1.5	105.8 ± 2.2	149.8 ± 1.9
100, chamber	155.3 ± 1.8	103.0 ± 2.1	154.3 ± 0.9
50, ambient	155.1 ± 2.1	117.3 ± 1.6	154.2 ± 1.6
50, chamber	154.2 ± 2.1	110.8 ± 2.2	155.1 ± 1.2
25, ambient	155.8 ± 1.8	118.4 ± 1.9	156.3 ± 1.8
25, chamber	156.3 ± 1.6	112.5 ± 1.8	156.1 ± 1.5

3.3. Electrochemical Measurements

We evaluated potentiodynamic behavior of laser-textured uncoated, epoxy-, and FAS-TiO_2/epoxy-coated stainless steel processed with scan line separation $x = y = 100$, 50, and 25 m in a simulated physiological Hank´s solution measured immediately after preparation, after three months, and one year when stored in ambient conditions (Figure 7). The polarization and the passivation behavior of the tested materials were significantly affected by the surface modification.

As shown in Figure 7a,d,g, immediately after preparation, the epoxy-, and FAS-TiO_2/epoxy-coated surfaces exhibited improved corrosion resistance with considerably broader range of passivation, which was also shifted to the lower corrosion–current densities, compared to uncoated surfaces. The observed shift was most pronounced for surfaces processed with $x = y = 25$ m. The corrosion parameters calculated from the potentiodynamic measurements indicated that the corrosion stability of coated specimens significantly increased, including lower corrosion–current densities (i_{corr}), lower corrosion rates (v_{corr}), and higher polarization resistances (R_p), compared to uncoated specimens (Table 4). The ratio between the smooth and the rough surface area, $A_0/A_{R,}$ listed in Table 1 was implemented in calculation of i_{corr} and v_{corr}. The electrochemical evaluation of surfaces exposed to the ambient conditions showed the enhanced short- and long-term corrosion stability of uncoated and coated laser-textured surfaces prepared under the same experimental conditions (Figure 7b,c,e,f,h,i).

Uncoated laser-textured surfaces, immediately after the laser texturing, that exhibited superhydrophilic nature (Table 4) showed similar corrosion resistance for all scanning line separations ($x = y = 100$, 50, and 25 m) (Table 4). This indicates that wetting properties dominate over the surface texture. After ageing, when the surfaces became (super)hydrophobic, we observed surface texture-dependent corrosion behavior. After three months, surfaces processed with scanning line separation $x = y = 50$ m showed better corrosion resistance compared to $x = y = 100$ m and 25 m.

As shown in Figure 1, surfaces processed with $x = y = 100$ m have a well-defined structure with a partly unprocessed steel surface, which is more susceptible to corrosion. On the other hand, the surfaces processed with $x = y = 25$ m were highly porous, which enabled the penetration of the aggressive ions, leading to the increased corrosion. This suggests that the surfaces processed with $x = y = 50$ m offer the best compromise between the laser texturing and, consequently, porosity. After one year, the corrosion resistance was further increased, since the increased long-term static contact angles reach superhydrophobic values on all the investigated uncoated laser-textured surfaces (Table 4). As the contact angles converge to similar values, we would expect similar corrosion resistance for all the aged uncoated laser-textured surfaces. However, corrosion resistance of $x = y = 25$ m surface remained lower due to high porosity, compared to well defined $x = y = 50$ and 100 m surfaces, which exhibited comparable corrosion stability. Observed long-term corrosion resistance of uncoated laser-textured surfaces is, therefore, effected not only by wetting properties of the surfaces but also by their morphology and imperfections, as reported by Boinovich et al. [20].

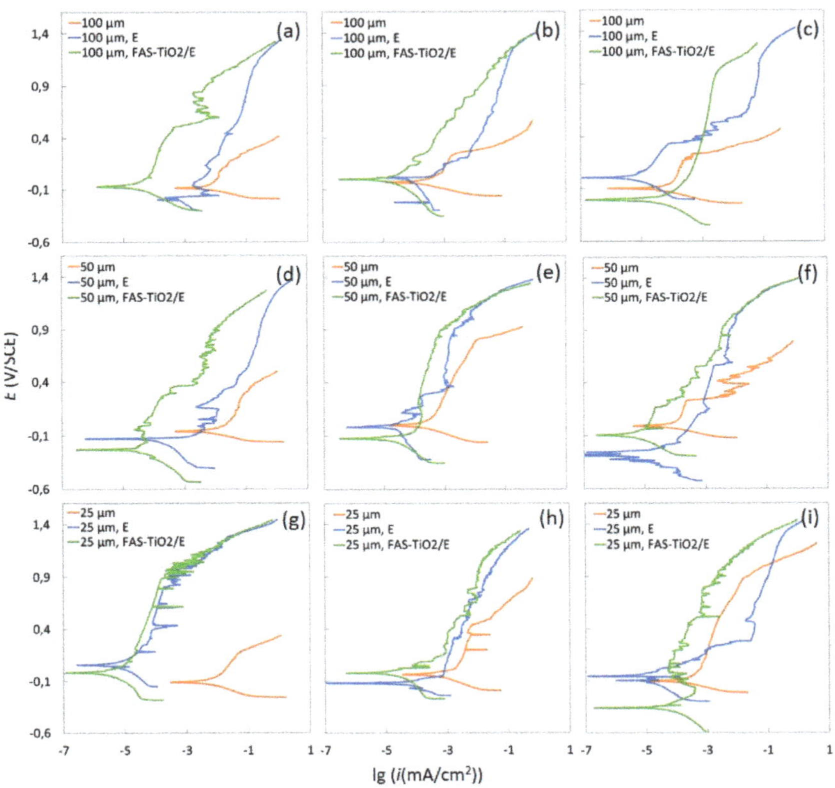

Figure 7. Potentiodynamic curves for uncoated, epoxy-coated, and FAS-TiO_2/epoxy-coated laser-textured stainless steel with different scan line separations ($\Delta x = \Delta y = 100, 50, 25$ µm) in a simulated physiological Hank´s solution measured (**a**,**d**,**g**) immediately after laser-texturing, (**b**,**e**,**h**) after 3 months, and (**c**,**f**,**i**) after 1 year storage in ambient conditions.

Table 4. Corrosion characteristics of uncoated, epoxy-coated and FAS-TiO_2/epoxy-coated laser-textured steel surfaces with different scan line separations ($\Delta x = \Delta y$ = 100, 50, 25 µm) calculated from the electrochemical measurements in a simulated physiological Hank´s solution and corresponding static water contact angles immediately after laser-texturing, after 3 months, and 1 year, when stored in ambient conditions.

Scan Line Separation [µm]	E_{corr} [mV]	i_{corr} [µA/cm^2]	v_{corr} [µm/year]	Contact Angle [°]
Immediately after preparation				
100 µm	−86.6 ± 0.5	4.345 ± 0.008	16.25 ± 0.07	0
100 µm, E	−172.9 ± 0.6	2.500 ± 0.006	3.45 ± 0.05	97.5 ± 1.4
100 µm, FAS-TiO_2/E	−70.7 ± 0.4	0.010 ± 0.001	0.25 ± 0.04	143.4 ± 3.1
50 µm	−59.7 ± 0.3	2.402 ± 0.005	20.51 ± 0.08	0
50 µm, E	−127.7 ± 0.5	0.209 ± 0.004	0.58 ± 0.03	112.1 ± 1.9
50 µm, FAS-TiO_2/E	−230.4 ± 0.9	0.007 ± 0.001	0.07 ± 0.01	151.1 ± 2.1
25 µm	−108.8 ± 0.5	5.540 ± 0.009	39.73 ± 0.07	0
25 µm, E	55.9 ± 0.2	0.013 ± 0.001	0.13 ± 0.01	108.3 ± 2.3
25 µm, FAS-TiO_2/E	−17.6 ± 0.1	0.002 ± 0.001	0.07 ± 0.01	151.9 ± 0.7
3 months after preparation				
100 µm	−30.6 ± 0.2	0.130 ± 0.007	0.70 ± 0.03	130.5 ± 0.6
100 µm, E	0.6 ± 0.1	0.155 ± 0.008	1.20 ± 0.04	104.3 ± 2.6
100 µm, FAS-TiO_2/E	−4.6 ± 0.1	0.005 ± 0.001	0.15 ± 0.01	140.5 ± 2.2
50 µm	1.6 ± 0.1	0.040 ± 0.004	0.42 ± 0.02	146.3 ± 1.6
50 µm, E	−11.1 ± 0.1	0.005 ± 0.001	0.12 ± 0.01	115.0 ± 2.2
50 µm, FAS-TiO_2/E	−121.4 ± 0.9	0.005 ± 0.001	0.05 ± 0.01	151.1 ± 0.9
25 µm	−35.7 ± 0.2	0.367 ± 0.006	2.60 ± 0.05	150.7 ± 1.1
25 µm, E	−120.3 ± 0.7	0.067 ± 0.003	0.40 ± 0.02	115.0 ± 1.2
25 µm, FAS-TiO_2/E	−20.4 ± 0.1	0.013 ± 0.002	0.27 ± 0.03	153.7 ± 1.3
1 year after preparation				
100 µm	−105.0 ± 0.9	0.015 ± 0.002	0.15 ± 0.01	155.3 ± 1.8
100 µm, E	−4.9 ± 0.8	0.005 ± 0.001	0.05 ± 0.01	103.0 ± 2.1
100 µm, FAS-TiO_2/E	−210.7 ± 0.9	0.035 ± 0.004	0.15 ± 0.01	154.3 ± 0.9
50 µm	−8.4 ± 0.1	0.009 ± 0.001	0.09 ± 0.01	154.2 ± 2.1
50 µm, E	−298.8 ± 0.3	0.009 ± 0.001	0.12 ± 0.01	110.8 ± 2.2
50 µm, FAS-TiO_2/E	−96.6 ± 0.5	0.002 ± 0.001	0.02 ± 0.01	155.1 ± 1.2
25 µm	−122.5 ± 0.6	0.127 ± 0.009	1.47 ± 0.02	156.3 ± 1.6
25 µm, E	−50.8 ± 0.9	0.007 ± 0.001	0.07 ± 0.01	112.5 ± 1.8
25 µm, FAS-TiO_2/E	−366.4 ± 0.3	0.060 ± 0.005	0.53 ± 0.05	156.1 ± 1.5

Corrosion resistance of coated surfaces immediately after preparation was, on the other hand, a combination of both wettability and surface morphology. We observed increased corrosion stability with decreased scanning line separation and the highest corrosion resistance was observed for FAS-TiO_2/epoxy-coated surface with $x = y$ = 25 and 50 m, which coincides with increased superhydrophobicity. After three months and one year, surfaces processed with $x = y$ = 100 m and 50 m and coated with E and FAS-TiO_2/E showed improved corrosion resistance compared to the samples immediately after the preparation. This suggests the superior role of protective coatings over the laser texturing and wettability. The contact angles of E-coated surfaces were slightly increased with time and remained moderately hydrophobic, whereas the wettability of FAS-TiO_2/E remained practically unchanged, i.e., superhydrophobic. The corrosion resistance for coated $x = y = 25$ m surfaces was, however, mainly decreased. This indicates that its highly porous structure leads to possible penetration of chloride ions to the metal surface under the coating, which increases the corrosion. Due to very stable wettability on a short- and long-term basis (Table 4), stable, long-term corrosion resistance can be, in general, attributed to long-term durability of coatings, which maintain high barrier properties. However, in some cases discrepancies from this trend were observed due to the coatings' imperfections.

4. Conclusions

Nanosecond laser texturing of austenitic stainless steel surfaces followed by surface modification with organic coating was used to fabricate surfaces with different surface characteristics, morphology, and wettability. Surface morphology was regulated by variation of the scanning line separation, whereas short-term and long-term wettability and corrosion resistance was evaluated and compared for uncoated laser-textured steel surfaces, epoxy-coated, and FAS-TiO_2/epoxy-coated textured steel surfaces in two storage media, i.e., in ambient conditions and in a chamber with reduced air pressure and humidity. With wettability evolution, we have shown two successful ways of creating a superhydrophobic surface: (i) indirectly by controllable ageing of the uncoated laser-textured steel surfaces in two different atmospheres, ambient conditions, and in a chamber with reduced air pressure and humidity; and (ii) directly by application of epoxy/FAS-TiO_2 coatings on laser-textured steel surfaces in both storage media. Storage conditions, air vs. reduced air pressure, and humidity strongly influence the wetting characteristics and wettability development of uncoated laser-textured steel. Reduced air pressure and humidity slow down the ageing process and superhydrophilic to superhydrophobic wettability transition. XPS analysis revealed the nature of the wettability conversion of uncoated laser-textured steel by adsorption of organic matters from surrounding media. Coated laser-textured samples, on the other hand, retained wetting properties over time in both media, leading to long-term stability of coatings with predictable wettability. Corrosion stability of uncoated laser-textured surfaces became laser pattern-regulated after wettability conversion. In general, stability was enhanced over time due to surfaces' superhydrophobicity. However, it was regulated also by their morphology and imperfections. The results show that coatings present a stable and effective long-term barrier in aggressive media and have a superior role over texture and wettability.

Author Contributions: Conceptualization, M.C. and P.G.; Methodology, M.C. and A.K.; Formal Analysis, M.C., A.K. and T.S.; Investigation, M.C. and T.S.; Writing—Original Draft Preparation, M.C. and A.K.; Writing—Review & Editing, M.C., A.K. and P.G.

Funding: This research was funded by the Slovenian Research Agency (research core funding Nos. P2-0132 and P2-0392 and project Nos. J2-7196 and J2-1741).

Acknowledgments: The authors acknowledge the support of SPI Lasers Ltd. by loan of their fiber laser within the research project Surface functionalization by nanosecond fiber laser texturing (nsFLaT).

Conflicts of Interest: The authors declare no conflict of interest.

References

1. Hryniewicz, T.; Rokicki, R.; Rokosz, K. Corrosion characteristics of medical-grade AISI Type 316L stainless steel surface after electropolishing in a magnetic field. *Corrosion* **2008**, *64*, 660–665. [CrossRef]
2. Ibrahim, M.A.M.; el Rehim, S.S.A.; Hamza, M.M. Corrosion behavior of some austenitic stainless steels in chloride environments. *Mater. Chem. Phys.* **2009**, *115*, 80–85. [CrossRef]
3. Long, J.Y.; Zhong, M.L.; Zhang, H.J.; Fan, P.X. Superhydrophilicity to superhydrophobicity transition of picosecond laser microstructured aluminum in ambient air. *J. Colloid Interface Sci.* **2015**, *441*, 1–9. [CrossRef] [PubMed]

4. Drzmala, J. Hydrophobicity and collectorless flotation of inorganic materials. *Adv. Colloid Interface Sci.* **1994**, *50*, 143–185. [CrossRef]
5. Long, J.Y.; Zhong, M.L.; Fan, P.X.; Gong, D.W.; Zhang, H.J. Wettability conversion of ultrafast laser structured copper surface. *J. Laser Appl.* **2015**, *27*, S29107. [CrossRef]
6. Kamegawa, T.; Irikawa, K.; Yamashita, H. Multifunctional surface designed by nanocomposite coating of polytetrafluoroethylene and TiO_2 photocatalyst: Self-cleaning and superhydrophobicity. *Sci. Rep.* **2017**, *7*, 13628. [CrossRef] [PubMed]
7. Xu, L.; Karunakaran, R.G.; Guo, J.; Yang, S. Transparent, superhydrophobic surfaces from one-step spin coating of hydrophobic nanoparticles. *ACS Appl. Mater. Interfaces* **2012**, *4*, 1118–1125. [CrossRef]
8. Zhang, L.; Zhao, N.; Xu, J. Fabrication and application of superhydrophilic surfaces: A review. *J. Adhes. Sci. Technol.* **2014**, *28*, 769–790. [CrossRef]
9. Zhang, X.; Jin, M.; Liu, Z.; Nishimoto, S.; Saito, H.; Murakami, T.; Fujishima, A. Preparation and photocatalytic wettability conversion of TiO_2-based superhydrophobic surfaces. *Langmuir* **2006**, *22*, 9477–9479. [CrossRef]
10. Lee, S.; Kim, W.; Yong, K. Overcoming the water vulnerability of electronic devices: A highly water-resistant ZnO nanodevice with multifunctionality. *Adv. Mater.* **2011**, *23*, 4398. [CrossRef]
11. Qing, Y.; Yang, C.; Sun, Y.; Zheng, Y.; Wang, X.; Shang, Y.; Wang, L.; Liu, C. Facile fabrication of superhydrophobic surfaces with corrosion resistance by nanocomposite coating of TiO_2 and polydimethylsiloxane. *Colloids Surf. A Physicochem. Eng. Asp.* **2015**, *484*, 471–477. [CrossRef]
12. Montemor, M.F. Functional and smart coatings for corrosion protection: A review of recent advances. *Surf. Coat. Technol.* **2014**, *258*, 17–37. [CrossRef]
13. Boinovich, L.B.; Emelyanenko, A.M.; Modestov, A.D.; Domantovsky, A.G.; Emelyanenko, K.A. Synergistic effect of superhydrophobicity and oxidized layers on corrosion resistance of aluminum alloy surface textured by nanosecond laser treatment. *ACS Appl. Mater. Interfaces* **2015**, *7*, 19500–19508. [CrossRef] [PubMed]
14. Ta, V.D.; Dunn, A.; Wasley, T.J.; Li, J.; Kay, R.W.; Stringer, J.; Smith, P.J.; Esenturk, E.; Connaughton, C.; Shephard, J.D. Laser textured superhydrophobic surfaces and their applications for homogeneous spot deposition. *Appl. Surf. Sci.* **2016**, *365*, 153–159. [CrossRef]
15. Gregorcic, P.; Setina-Batic, B.; Hocevar, M. Controlling the stainless steel surface wettability by nanosecond direct laser texturing at high fluences. *Appl. Phys. A Mater. Sci. Process.* **2017**, *123*, 766. [CrossRef]
16. Ta, V.D.; Dunn, A.; Wasley, T.J.; Li, J.; Kay, R.W.; Stringer, J.; Smith, P.J.; Esenturk, E.; Connaughton, C.; Shephard, J.D. Laser textured surface gradients. *Appl. Surf. Sci.* **2016**, *371*, 583–589. [CrossRef]
17. Ta, D.V.; Dunn, A.; Wasley, T.J.; Kay, R.W.; Stringer, J.; Smith, P.J.; Connaughton, C.; Shephard, J.D. Nanosecond laser textured superhydrophobic metallic surfaces and their chemical sensing applications. *Appl. Surf. Sci.* **2015**, *357*, 248–254. [CrossRef]
18. Kietzig, A.M.; Hatzikiriakos, S.G.; Englezos, P. Patterned Superhydrophobic Metallic Surfaces. *Langmuir* **2009**, *25*, 4821–4827. [CrossRef]
19. Gregorčič, P.; Conradi, M.; Hribar, L.; Hočevar, M.; Gregorcic, P.; Conradi, M.; Hribar, L.; Hocevar, M. Long-term influence of laser-processing parameters on (super)hydrophobicity development and stability of stainless-steel surfaces. *Materials* **2018**, *11*, 2240. [CrossRef]
20. Boinovich, L.B.; Gnedenkov, S.V.; Alpysbaeva, D.A.; Egorkin, V.S.; Emelyanenko, A.M.; Sinebryukhov, S.L.; Zaretskaya, A.K. Corrosion resistance of composite coatings on low-carbon steel containing hydrophobic and superhydrophobic layers in combination with oxide sublayers. *Corros. Sci.* **2012**, *55*, 238–245. [CrossRef]
21. Trdan, U.; Hocevar, M.; Gregorcic, P. Transition from superhydrophilic to superhydrophobic state of laser textured stainless steel surface and its effect on corrosion resistance. *Corros. Sci.* **2017**, *123*, 21–26. [CrossRef]
22. Emelyanenko, A.M.; Shagieva, F.M.; Domantovsky, A.G.; Boinovich, L.B. Nanosecond laser micro- and nanotexturing for the design of a superhydrophobic coating robust against long-term contact with water, cavitation, and abrasion. *Appl. Surf. Sci.* **2015**, *332*, 513–517. [CrossRef]
23. Fairley, N. *Casa XPS VAMAS Processing Software*; Casa Software Ltd.: Teignmouth, UK, 2018.
24. Conradi, M.; Drnovšek, A.; Gregorčič, P. Wettability and friction control of a stainless steel surface by combining nanosecond laser texturing and adsorption of superhydrophobic nanosilica particles. *Sci. Rep.* **2018**, *8*, 7457. [CrossRef] [PubMed]
25. McHale, G.; Shirtcliffe, N.J.; Newton, M.I. Super-hydrophobic and super-wetting surfaces: Analytical potential? *Analyst* **2004**, *129*, 284–287. [CrossRef]

26. Wang, Z.K.; Zheng, H.Y.; Lim, C.P.; Lam, Y.C. Polymer hydrophilicity and hydrophobicity induced by femtosecond laser direct irradiation. *Appl. Phys. Lett.* **2009**, *95*, 111110. [CrossRef]
27. Lai, J.N.; Sunderland, B.; Xue, J.M.; Yan, S.; Zhao, W.J.; Folkard, M.; Michael, B.D.; Wang, Y.G. Study on hydrophilicity of polymer surfaces improved by plasma treatment. *Appl. Surf. Sci.* **2006**, *252*, 3375–3379. [CrossRef]
28. Yang, Z.; Liu, X.; Tian, Y. Insights into the wettability transition of nanosecond laser ablated surface under ambient air exposure. *J. Colloid Interface Sci.* **2019**, *533*, 268–277. [CrossRef]

coatings

MDPI

Article

Bioactive Coating on Titanium Dental Implants for Improved Anticorrosion Protection: A Combined Experimental and Theoretical Study

Jozefina Katić [1], Ankica Šarić [2,*], Ines Despotović [3], Nives Matijaković [4], Marin Petković [5] and Željka Petrović [4,*]

1 Department of Electrochemistry, Faculty of Chemical Engineering and Technology, University of Zagreb, Marulićev trg 19, 10000 Zagreb, Croatia; jkatic@fkit.hr
2 Division of Materials Physics, Centre of Excellence for Advanced Materials and Sensing Device, Ruđer Bošković Institute, Bijenička cesta 54, 10002 Zagreb, Croatia
3 Division of Physical Chemistry, Ruđer Bošković Institute, Bijenička cesta 54, 10002 Zagreb, Croatia; Ines.Despotovic@irb.hr
4 Division of Materials Chemistry, Ruđer Bošković Institute, Bijenička cesta 54, 10002 Zagreb, Croatia; Nives.Matijakovic@irb.hr
5 Adentro dental studio, Petrova ul. 67, 10000 Zagreb, Croatia; info@adentro.hr
* Correspondence: Ankica.Saric@irb.hr (A.Š.); Zeljka.Petrovic@irb.hr (Ž.P.)

Received: 26 July 2019; Accepted: 23 September 2019; Published: 25 September 2019

Abstract: In recent years, extensive studies have been continuously undertaken on the design of bioactive and biomimetic dental implant surfaces due to the need for improvement of the implant–bone interface properties. In this paper, the titanium dental implant surface was modified by bioactive vitamin D3 molecules by a self-assembly process in order to form an improved anticorrosion coating. Surface characterization of the modified implant was performed by field emission scanning electron microscopy (FE-SEM), attenuated total reflection Fourier transform infrared spectroscopy (ATR-FTIR), and contact angle measurements (CA). The implant's electrochemical stability during exposure to an artificial saliva solution was monitored in situ by electrochemical impedance spectroscopy (EIS). The experimental results obtained were corroborated by means of quantum chemical calculations at the density functional theory level (DFT). The formation mechanism of the coating onto the titanium implant surface was proposed. During a prolonged immersion period, the bioactive coating effectively prevented a corrosive attack on the underlying titanium (polarization resistance in order of 10^7 Ω cm^2) with ~95% protection effectiveness.

Keywords: titanium dental implant; vitamin D3; bioactive coating; anticorrosion protection; electrochemical impedance spectroscopy (EIS); density functional theory level (DFT)

1. Introduction

Titanium and its alloys are the most commonly used implant materials, which represent gold standards for dental implant fabrication due to their exceptional combination of high corrosion resistivity, favourable mechanical properties, and biocompatibility [1–3]. Although the implantation success of titanium dental implants is high, complications still occur. There is an increasing number of papers regarding titanium hypersensitivity, skin allergies such as contact dermatitis, eczema, and immune reactions appearing after implant fixation [4–7]. It is also known that osseointegration problems in patients suffering from osteoporosis or other bone-related problems are one of reasons for unsuccessful implantation of dental implants [8]. Therefore, there is a need for continuous improvement of dental implant material characteristics as well as their design to ensure long-term integration of implants into the jaw.

Surface modification, which in recent years has been in the focus of extensive research, is one of strategies for improving corrosion resistivity, biocompatibility, and bioactivity of dental implants [3,8–10]. It has been reported that the presence of inorganic coatings such as calcium phosphates [11] or hydroxyapatite [12,13] promotes and accelerates bone formation in implant surroundings. Nano-hydroxyapatite is used as a single coating or in a combination with collagen, bio glass, or titanium dioxide in a composite coating to imitate the bio-environment of native bones [14]. Functionalization of the implant surfaces with organic or bioactive molecules promotes the adhesion of various cells [3]. Coatings based on bisphoshonates—drugs for bone disease treatments—act osseoinductively and improve bone–implant connections [8,15,16].

According to currently available data, there is a great interest among dental surgeons and implantologists in analyzing relationships between vitamin D3 and the osseointegration process [17,18]. Besides a known role of vitamin D3 in bone metabolism, D3 is also very relevant for the normal functioning of the immune system, which is of particular importance for a successful integration of a dental implant in surrounding bone tissue [17]. Therefore, in this study, the titanium dental implant surface was modified by a bioactive vitamin D3 coating prepared by a self-assembly process. The main goal was (i) to make the dental implant surface more osseoinductive and simultaneously (ii) more corrosion resistant during exposure to aggressive media (oral cavity fluids). Based on the results of a combined electrochemical–theoretical study, the bioactive coating formation mechanism was clarified as well as the coating's influence on electrochemical behavior of the titanium implant during immersion in an artificial saliva solution. To the best of our knowledge, the present study shows for the first time the results of an integrated theoretical–experimental approach, which contribute to a fundamental understanding of bioactive and protective coating formation and predict overall electrochemical stability of modified dental implants in oral cavity fluids.

2. Materials and Methods

2.1. Material and Chemicals

The Ankylos® C/X training implant A11 (length: 11 mm, diameter: 3.5 mm, Dentsply Friadent GmbH, Mannheim, Germany), denoted as Ti-implant, was the object of the study and it was used as-received. According to manufacturer's data, the implant is made of pure titanium grade 2, ISO 5832-2:2018 [19], and its chemical composition is presented in Table 1 [20].

Table 1. Chemical composition (wt %) of titanium c.p. grade 2 [20].

Element	N	C	O	Fe	H	Ti	Other
wt %	0.03	0.10	0.25	0.30	0.0155	Balance	0.4

Vitamin D3 drops (ChildLife®, Los Angeles, CA, USA) were used as-received. 122 μmol dm^{-3} cholecalciferol (vitamin D3) in aqueous glycerol solution was used for modification of the Ti-implant surface. The chemical structure of the vitamin D3, $C_{27}H_{44}O$, is presented in Figure 1.

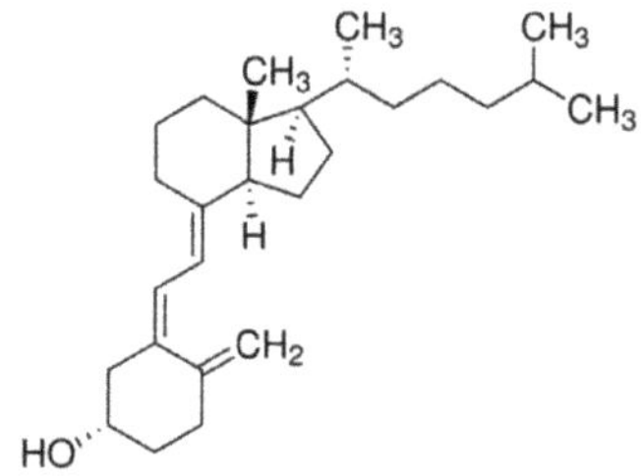

Figure 1. The chemical structure of the vitamin D3.

2.2. Formation of Bioactive Coating

Before modification, the surface of as-received Ti-implant was ultrasonically cleaned with acetone (p.a., Gram-Mol, Zagreb, Croatia) and redistilled water, degreased in absolute ethanol (p.a., Gram-Mol, Croatia), and dried in a stream of nitrogen (99.999%, Messer, Bad Soden, Germany). The D3 vitamin layer onto the Ti-implant surface was prepared by a self-assembly method. The Ti-implant was immersed in an aqueous glycerol solution of vitamin D3 at 24 ± 2 °C during 24 h. Afterwards, the coated Ti-implant was dried in a regular air-convection oven (Instrumentaria, Sesvete, Croatia) at 70 °C for 5 h. That well-known procedure [21,22] enhanced the adhesion and stability of the coating due to the conversion of the hydrogen-bonded intermediate into the stable coating with a chemical Ti–O bond by an acid-base condensation mechanism. The formation mechanism as well as the type of interactions were accurately corroborated by DFT (Section 3.3). Finally, the modified sample was rinsed with absolute ethanol, redistilled water, and dried in a stream of nitrogen. The prepared modified sample, denoted in the text as the Ti-implant/bioactive coating, was used for further characterization.

2.3. Characterisation Methods

The morphology characteristics of unmodified and modified Ti-implant surfaces were examined by a thermal field emission scanning electron microscope (model JSM-7000F, Jeol Ltd., Tokyo, Japan) at 15 kV.

The contact angles on investigated Ti-implant surfaces were measured with a drop of 1 μL Milli-Q® water (Milli-Q® Direct 8 Water Purification System, Merck, Darmstadt, Germany) at ambient atmospheric conditions using a contact angle system OCA 20 (Dataphysics Instruments GmbH, Filderstadt, Germany). Values reported are the average of three measurements taken at smooth upper part of the implant (without threads) after initial period of 10 s stabilization.

The ATR-FTIR measurements were performed by a Tensor II spectrometer (Bruker Optik GmbH, Ettlingen, Germany) over the range of 4000–340 at 4 cm^{-1} scan step and total 16 scans per measurement.

Electrochemical characterization of unmodified and modified Ti-implant samples was assessed by electrochemical impedance spectroscopy (EIS). A standard three-electrode cell (Corrosion Cell 6.1415.250, volume 50–150 mL, Metrohm Autolab B.V., Utrecht, Netherlands) was utilized with the Ti-implant material as a working electrode (an area of 0.98 cm^2 exposed to the electrolyte solution). A large-area Pt electrode served as a counter electrode and a reference electrode, to which all potentials in the paper are referred, was an Ag|AgCl, 3.0 mol dm^{-3} KCl (E = 0.210 V vs. standard hydrogen electrode, SHE). Barrier properties of unmodified and modified Ti-implants were evaluated in solution based on Fusayama artificial saliva (0.4 g dm^{-3} NaCl, 0.4 g dm^{-3} KCl, 0.6 g dm^{-3} $CaCl_2 \cdot 2H_2O$, 0.58 g dm^{-3} $Na_2HPO_4 \cdot 2H_2O$, and 1 g dm^{-3} urea), pH 6.8 [23], prepared from p.a. grade chemicals and redistilled water. Prior to EIS measurements, the Ti-implant electrode was stabilized for 1 h, 1, and 7 days at the open circuit potential, E_{OC}, in the electrolyte solution.

EIS measurements were performed subsequently in the frequency range from 10^5 to 10^{-3} Hz at E_{OC} with an *ac* voltage amplitude of ±5 mV using Solartron 1287 potentiostat/galvanostat with Solartron FRA 1260 (Solartron Analytical, Farnborough, UK) controlled by ZPlot® software (v. 3.5e, Southern Pines, Moore, NC, USA). The complex non-linear least squares (CNLS) fit analysis software [24] was employed to model the experimental data obtained. The values of the elements of the proposed electric equivalent circuit (EEC) were derived with χ^2 values less than 5×10^{-3} (errors in parameter values of 1–3%) using ZView® software (v. 3.5e, Southern Pines, USA).

2.4. Computational Details

All calculations were performed by means of quantum chemical calculations at the density functional theory (DFT) level using the Gaussian 09 program (revision D1) [25]. The small $(TiO_2)_{10}$ nanocluster served as a credible model for all possible molecular surface/bioactive molecule interaction predictions [26,27]. The M06 functional designed by Truhlar's group was selected [28–30]. For the geometry optimization the 6-31+G(d,p) + LANL2DZ basis set was utilized, which means that the Pople's 6-31+G(d,p) double-ξ basis set was chosen for O, H, C atoms and the LANL2DZ basis for the transition-metal (Ti) atoms [31]. Frequency calculations were done under the harmonic approximation on all the optimized structures at the same level of theory, with no scaling in order to confirm the true minima of the structures. The final single point energies were obtained using a highly flexible 6-311++G(2df,2pd) basis set for the O, H, C atoms, while the same LANL2DZ ECP type basis set for titanium atoms was employed. To evaluate the bulk solvent effects (1,2-ethandiol as a glycerol approximation, $\varepsilon = 40.245$), the implicit SMD polarizable continuum solvation model [32] was employed.

The Gibbs free energy interactions, ΔG^*_{INT} were computed as the difference between the total free energy (G^*_{AB}) of the resulting (AB) structure and the sum of the total free energies ($G^*_A + G^*_B$) of the associating units A and B (Tables S1–S4 in Supplementary Materials).

The topological analysis of the charge density distribution was performed by employing AIMALL software package [33] using Bader's quantum theory of atoms in molecules (QTAIM) [34] with SMD/M06/6-31+G(d,p) + LANL2DZ wave function derived from the optimization. A detailed description of the computational modelling is given in the Supplementary Material.

3. Results and Discussion

3.1. Surface Characterization of the Unmodified and Modified Ti-Implant Samples

The surface morphology of both unmodified (as-received Ti-implant) and modified (Ti-implant/bioactive coating) samples was investigated by using FE-SEM and is presented in Figure 2. SEM images of the Ti-implant surface present a inhomogeneous microstructured surface layer exhibiting different size cavities and defects and is characterized by overall microroughness (Figure 2a,b). According to the manufacturer's data, the implant surface was grit-blasted and high-temperature etched and this microstructure is commercially known as Friadent® plus microstructure. The resulting surface favors the highly rapid apposition of bone-inducing cells on the implant [19]. Observed morphology is in accordance with SEM results reported by R. Smeets et al. [8], F. Rupp et al. [35], and T.J. Webster et al. [36]. The surface morphology did not change significantly upon the bioactive coating on the Ti-implant surface preparation. Since the D3 layer thickness was about 0.6 nm, according to the DFT results, the level of possible morphology differentiation by SEM was very low.

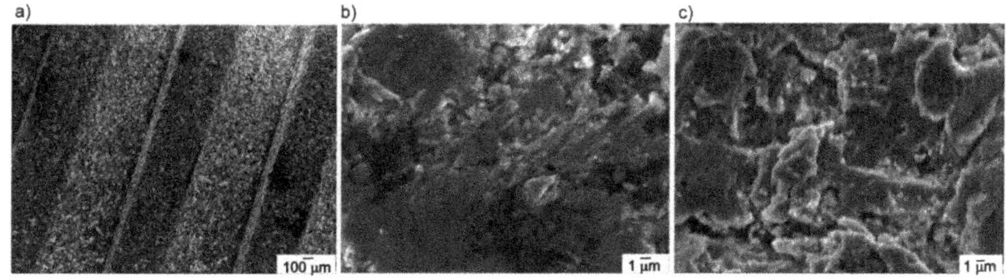

Figure 2. SEM images of the unmodified Ti-implant surface at: (**a**) 50×; (**b**) 3000× and (**c**) of the Ti-implant/bioactive coating surface at 3000× magnification.

The surface chemistry of the unmodified (Ti-implant) and modified (Ti-implant/bioactive coating) samples was examined by ATR-FTIR and recorded spectra are presented in Figure 3a. The spectrum of the Ti-implant/bioactive coating sample was recorded immediately after the final step of the coating formation (see Section 2.2). A confirmation of a successful formation of the bioactive coating on the Ti-implant surface was deduced based on the presence of vitamin D3 functional group bands in investigated samples spectra, which indicates chemisorption on the Ti-implant surface, as will be given below.

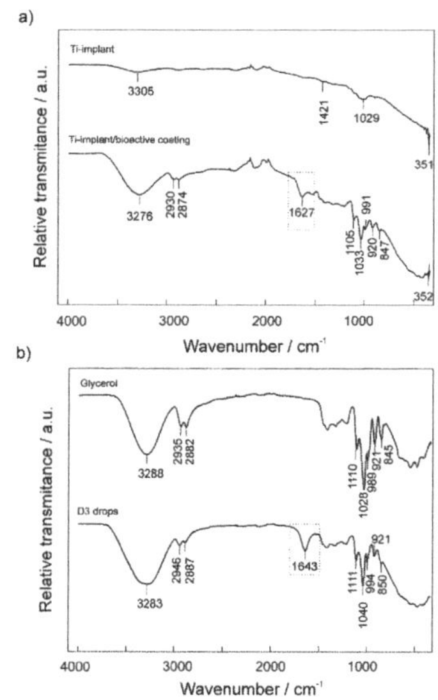

Figure 3. Attenuated total reflection Fourier transform infrared spectroscopy (ATR-FTIR) spectra of (**a**) the unmodified and modified Ti-implant surfaces and (**b**) D3 drops and glycerol.

In the Ti-implant sample ATR-FTIR spectrum, a broad band observed at 3305 cm^{-1} is characteristic to the stretching vibration of adsorbed water on the Ti-implant surface. A vibration band characteristic for the Ti–OH appeared at 1029 cm^{-1}. A band at 1421 cm^{-1} arising from TiO_2 lattice vibration and the bands located in the range from 350–1000 cm^{-1} were related to the Ti–O stretching vibrations [37,38]. According to the described bands, the Ti-implant surface is covered by a TiO_2 layer.

In Figure 3a a spectrum of the Ti-implant/bioactive coating sample is presented. ATR-FTIR spectra of vitamin D3 and glycerol were also recorded since vitamin D3 drops used for the bioactive coating formation contained glycerol as a solvent (Figure 3b). As can be seen, spectra of both compounds are very similar and are in accordance with the literature data [39]. A clear difference is a peak at 1643 cm^{-1} that is assigned to the H–C=CH stretching vibrations characteristic to vitamin D3.

A spectrum of the Ti-implant/bioactive coating (Figure 3a) contains observed distinctive bands for vitamin D3 and glycerol. A presence of broader band at 1627 cm^{-1}, characteristic for vitamin D3 molecule, confirms a successful formation of the vitamin D3 layer on the Ti-implant surface. Shifts of all bands to lower wavenumbers also indicate an attachment (chemisorption) of vitamin D3 molecules to the Ti-implant surface. A peak at 351 cm^{-1} of TiO_2 is visible due to relatively low D3 layer thickness, according to the DFT results.

Since the ATR-FTIR spectra of D3 and glycerol molecules exhibited high degree of resemblance and intermolecular interactions exist between both molecules (DFT results, Section 3.3), there is a possibility that both compounds are simultaneously bonded to the implant surface as a two layer-structured coating, as will be discussed later (Section 3.3). Therefore, it is impossible to accurately determine the interaction nature and type between vitamin D3 molecules and Ti-implant substrates only from the observed band shifts. Further experimental characterization as well as theoretical DFT calculations were carried out to accurately determine the structure of the bioactive coating formed onto the Ti-implant surface.

A study of the wetting properties of implant surfaces is very important for the understanding of a complex reaction series occurring during the initial contact between implant and body fluids [8,35]. In this study, wettability measurements were conducted in order to evaluate level of a bioactive coating formation success. A static contact angle of water, θ, was measured on the unmodified (Ti-implant) and modified Ti-implant (Ti-implant/bioactive coating) surfaces. Measurements were carried out at upper smooth part of the implant body and corresponding micrographs are presented in Figure 4.

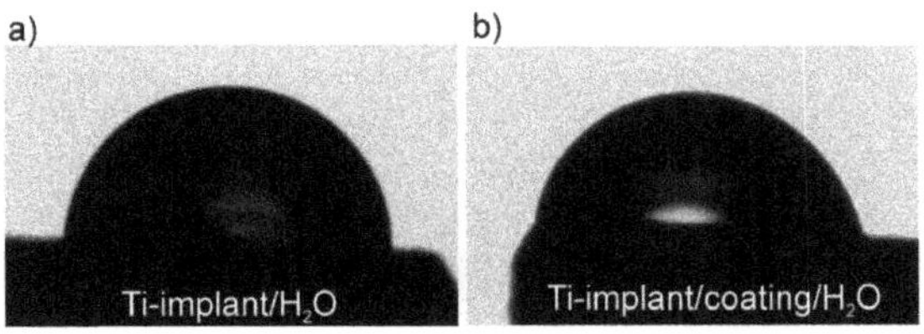

Figure 4. Optical micrographs of a water drop on (**a**) the Ti-implant surface and (**b**) the Ti-implant/bioactive coating surface.

The higher value of contact angle, $\theta = 87.5 \pm 2.2°$, was obtained on the Ti-implant surface in comparison to the value obtained on the Ti-implant/bioactive coating surface, $\theta = 60.4 \pm 1.5°$. Wetting properties of the Ti-implant surface were changed by a presence of a coating and these results were confirmation of a successful formation of the bioactive coating on the implant surface. The surface modification induced a change from a hydrophobic (Ti-implant) to hydrophilic surface character (Ti-implant/bioactive coating). Therefore, the wetting properties of the modified Ti-implant surface are predominantly affected by the hydrophilic –OH functional groups oriented in the upper part of the coating, possibly originating from glycerol molecules, as will be discussed on the basis of DFT results (see Section 3.3).

3.2. Electrochemical Characterization of the Unmodified and Modified Ti-Implant Samples

The electrochemical behavior of the unmodified and D3 vitamin modified Ti-implant surfaces, as one of prerequisite factors for dental implant biocompatibility in contact with aggressive oral cavity fluids and hence successful implantation process, was studied under in vitro conditions of real dental implant application. Employing a non-destructive method of electrochemical impedance spectroscopy (EIS), electrochemical stability of investigated samples during different immersion times (shown in Figure 5) was monitored in situ in an artificial saliva electrolyte solution at E_{OC} over the wide frequency range and the obtained impedance spectra are given in the form of Bode magnitude and phase angle plots in Figure 5a,b. The impedance data were analyzed in terms of the electric equivalent circuits

(EECs) that are presented in Figure 5c,d and the impedance parameter values are provided in Tables 2 and 3.

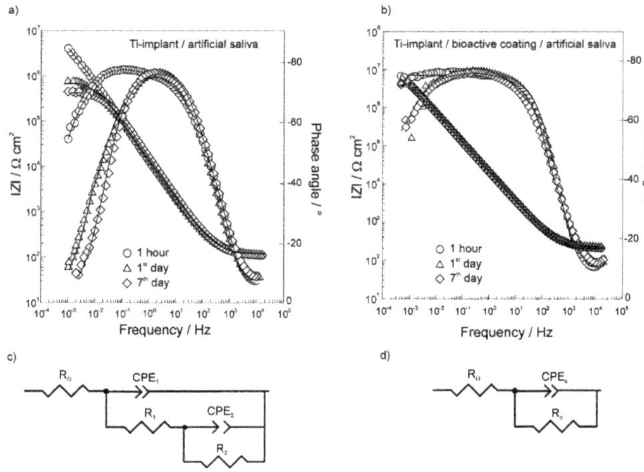

Figure 5. The Bode plots of (**a**) the unmodified Ti-implant and (**b**) the modified Ti-implant recorded at E_{OC} in the artificial saliva electrolyte solution, pH 6.8 after various immersion times denoted. Symbols—the experimental data; solid lines—the modeled data. The schematic presentation of the electric equivalent circuits (EECs) used to fit the spectra of (**c**) the unmodified Ti-implant and (**d**) the modified Ti-implant.

Table 2. Impedance parameters calculated for electrochemical impedance spectroscopy (EIS) data of the Ti-implant/oxide film/artificial saliva interface (Figure 5a) recorded at E_{OC} after various exposure times denoted.

Exposure Time	R_Ω/Ω cm^2	$10^6 \times Q_1/\Omega^{-1}$ cm^{-2} s^n	n_1	C_1/μF cm^{-2}	R_1/Ω cm^2	$10^6 \times Q_2/\Omega^{-1}$ cm^{-2} s^n	n_2	C_2/μF cm^{-2}	R_2/MΩ cm^2
1 h	111	9.98	0.853	3.02	760	5.16	0.850	1.38	9.90
1 day	123	2.88	0.997	2.88	174	12.1	0.788	2.10	0.79
7 days	123	3.26	0.978	2.71	307	9.21	0.810	1.88	0.44

Table 3. Impedance parameters and protection effectiveness values calculated for EIS data of the Ti-implant/oxide film/D3 layer/artificial saliva interface (Figure 5b) recorded at E_{OC} after various exposure times denoted.

Exposure Time	R_Ω/Ω cm^2	$10^6 \times Q_c/\Omega^{-1}$ cm^{-2} s^n	n	C_c/μF cm^{-2}	R_c/MΩ cm^2	η/%
1 h	157	7.97	0.851	2.52	51.7	80.9
1 day	157	6.95	0.862	2.54	17.4	95.5
7 days	157	8.14	0.846	2.47	17.6	97.5

In the EECs employed, the constant-phase element, CPE, was used to describe a non-ideal capacitance behavior in order to compensate non-homogeneity in the system (inhomogeneous current flow, capacitance dispersion, etc.) [40–42]. The impedance of CPE can be expressed as $Z_{CPE} = [Q(j\omega)^n]^{-1}$ where Q and n are parameters associated with CPE. For CPE exponent $n = 1$, the frequency–independent CPE parameter Q represents the capacity of the interface. For $n \neq 1$, the system shows behavior that has been attributed to surface heterogeneity, the presence of surface films, or to continuously distributed time constants for charge–transfer reactions [40]. The corresponding interface capacitance values were calculated according to Brug's relation [41]:

$$Q = C^n(R_\Omega^{-1} + R^{-1})^{1-n} \quad (1)$$

The EIS data of the unmodified Ti-implant (Ti-implant/oxide film/artificial saliva interface) were fitted to the EEC with two time constants (Figure 5c) and the parameter values are presented in Table 2. The EEC used is commonly applied to analyze impedance results of Ti and Ti-alloys covered with a bi-layer structure of oxide film [22,43–48]. The high/middle frequency time constant (R_1-CPE$_1$) describes the characteristics of the outer porous layer of the oxide film. R_1 represents the resistance and CPE$_1$ the capacitance of the outer layer. The low frequency time constant (R_2-CPE$_2$) is related to the inner barrier layer of the oxide film, predominately containing titanium(IV) oxide [1,44,49]. R_2 represents the resistance and CPE$_2$ the capacitance of the barrier layer. R_Ω is the ohmic (electrolyte) resistance. The polarization (corrosion) resistance, R_p, is the sum of R_1 and R_2 resistance contributions and determines the overall corrosion resistance of the oxide-covered Ti-implant [50].

As can be seen from Figure 5a and Table 2, the electrochemical stability of the Ti-implant sample was significantly decreased during an exposure to an artificial saliva solution, which was particularly reflected in R_1 and R_2 values. During the initial short immersion period, an oxide film on the Ti-implant surface possessed high protective properties (high R_p value). Since the artificial saliva solution represents an aggressive medium, a high amount of chloride ions affects the oxide film properties and accelerates the titanium corrosion (degradation) process during a prolonged immersion period [51]. Calculated parameters, especially R_1 values, indicate that the contribution of the outer layer of the oxide film to the electrochemical behaviour is rather imperceptible, and the overall EIS response is dominated by the inner barrier layer of the oxide film, which is in accordance with previously reported results [43,44].

The Bode plots of the Ti-implant/bioactive coating sample measured after specified artificial saliva exposure times are presented in Figure 5b. The brief inspection of the spectra points to the different impedance behavior of vitamin D3 modified Ti-implant surface in comparison to the unmodified Ti-implant surface. Hence, EEC with only one time constant was employed to analyze the EIS data (Figure 5d), with R_c and C_c attributing as the resistance and capacitance of the bioactive coating formed. The bioactive coating structure is very compact and well-ordered due to the presence of vitamin D3 layer over the oxide covered Ti-implant surface, as was confirmed by the DFT calculation results (Section 3.3) and both of layers contribute to the overall corrosion resistance of the underlying Ti-implant. The resistance R_c represents the polarization resistance, R_p, of the investigated system. The fitted values are presented in Table 3.

As can be seen, the dependence of phase angle versus $\log f$ points to a capacitive behavior of the coated Ti-implant sample over the wider frequency range in comparison to the unmodified Ti-implant sample. Besides, higher low frequency magnitude values that remain almost unchanged during the investigated period of 7 days indicate improved electrochemical stability and corrosion protection of the Ti-implant upon bioactive coating formation.

However, deep insight into phase angle versus $\log f$ dependence, as a structural sensitive parameter, reflects possible structural changes inside the bioactive coating during first day of sample immersion into the artificial saliva solution. This rearrangement/reorganization inside the bioactive coating induced a slight deterioration of protective properties, as can be seen from R_c values (Table 3). After a prolonged immersion period (from 1st to 7th day), EIS responses did not change significantly. Obviously, observed coating rearrangement/reorganization was induced by an initial contact between modified Ti-implant surface and electrolyte solution. According to the literature, structural reorganization that modulates electrical properties is known for self-assembled surface layers and can be stimulated by potential or by polarity and wettability of the terminal functional group of the layer molecules [52,53]. On the basis of DFT calculations (Section 3.3), weak intermolecular interactions between glycerol and D3 molecules are present and they result in glycerol molecules orientation in the outer part of the bioactive coating. Thus, glycerol–OH terminal functional groups affect the polarity and wettability of the coating formed on the Ti-implant surface, which was observed in the contact angle value (Figure 4). Probably, there is a possibility of stronger interaction between the glycerol -OH group and electrolyte components that caused a glycerol withdrawal from the Ti-implant surface and simultaneous structural changes inside the bioactive coating, as was confirmed by DFT (Section 3.3). The coated Ti-implant surface was perturbed and small defects in the coating's structure probably occurred. Consequently, ions/water molecules could penetrate into the coating, hence causing a decrease in R_c values after a one-day immersion. During this first day, the coating structure again achieved the most stable conformation resulting in unchanged R_c values. It should be stressed that, albeit the R_p value decrease was observed due to the structural organization occurrence, the overall coating resistance (the sum of oxide film + D3 layer resistances) inversely proportional to the corrosion rate, remained in order of 10^7 Ω cm^2 and imparted sufficient corrosion protection to the underlying Ti-implant material.

The corrosion protection effectiveness, η, of the bioactive coating formed on the Ti-implant surface was calculated using the relation:

$$\eta = (R_{p\,(\text{modified})} - R_{p\,(\text{umodified})})/R_{p\,(\text{modified})} \quad (2)$$

where $R_{p\,(\text{unmodified})}$ and $R_{p\,(\text{modified})}$ are the polarization resistances of unmodified and modified Ti-implants. The protection effectiveness values are presented in Table 3.

According to EIS results, protective properties of the unmodified, as-received Ti-implant (Ti-implant/oxide film) were significantly deteriorated during its immersion into an artificial saliva solution. On the other hand, the D3 layer, prepared by self-assembly process on the oxide-covered Ti-implant surface (Ti-implant/bioactive coating), behaved as an excellent barrier to the transport of corrosive ions/molecules from the bulk electrolyte to the Ti-implant surface during a prolonged exposure. Additionally, vitamin D3 molecules are bioactive molecules that can promote and accelerate an osseointegration process. Thus this bioactive, and at the same time, highly protective coating can serve as a good candidate for biocompatible Ti dental implants, enabling a pivotal role in the successful implantation process.

3.3. *Formation Mechanism of Bioactive Implant Coating*

Notwithstanding the part of the formation mechanism that has been discussed above in considering experimental results, a more detailed theoretical study of Ti-implant surface/bioactive molecule interactions is still needed to fully clarify the coating formation mechanism. The experimental findings were corroborated by means of DFT calculations based on effects of cholecalciferol (D3 vitamin) as well as glycerol as its alcoholic solvent in pharmaceutical composition of vitamin D3 solution. The small $(TiO_2)_{10}$ nanocluster served as a credible model for all possible molecular Ti-implant surface/bioactive molecule interaction predictions [26,27]. There is a possibility that the presence of other kinds of Ti-implant surface/molecule interactions (due to the presence of glycerol solvent) except the $(TiO_2)_{10}$–cholecalciferol interaction could be responsible for influencing the coating formation mechanism. Hence, the implant surface/bioactive molecule interactions were investigated considering the results obtained from a computational study of Gibbs free energies of $(TiO_2)_{10}$–cholecalciferol, $(TiO_2)_{10}$–cholecalciferol–glycerol, as well as $(TiO_2)_{10}$–glycerol molecular interactions (ΔG^*_{INT}). One should mention that only the most thermodynamically stable structures are discussed here.

Different binding interactions were established, involving the processes of the $(TiO_2)_{10}$–cholecalciferol interaction. The interactions, which involve Ti–O bonding and C–H···O hydrogen bonding, are shown in Figure 6. When Ti–O and hydrogen bonds are formed, the free energies of $(TiO_2)_{10}$–cholecalciferol molecular interactions are released ($\Delta G^*_{INT} = -6.64$ kcal mol^{-1}). The $(TiO_2)_{10}$–cholecalciferol interaction has been found to be a spontaneous exergonic process. The formation of such a completely enclosed structure of $(TiO_2)_{10}$–cholecalciferol, with a high coverage level of the $(TiO_2)_{10}$ surface, is likely a consequence of the synergistic effect of all the interactions mentioned above. Namely, the most stable structure of $(TiO_2)_{10}$–cholecalciferol possesses a Ti–O bond ($d_{Ti–O}$ = 2.252 Å, $E_{Ti–O}$ = −14.68 kcal mol^{-1}) which is stabilized by eleven weaker C–H···O hydrogen bonds ($E_{O...H}$ ranges from −0.61 to −2.99 kcal mol^{-1}; $d_{O...H}$ ranges from 3.295 to 2.358 Å). The Ti–O bond is attributed to an ionic (polar coordinate) type of interacting, according to $\Delta^2\rho$ (r_c) > 0 and $H(r_c)$ > 0.

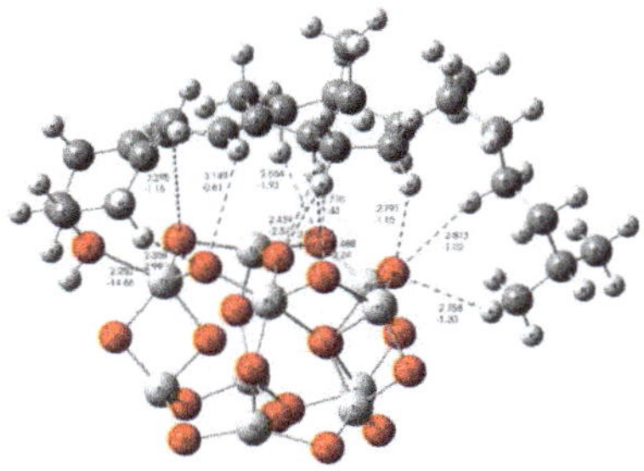

Figure 6. The most stable structure of the $(TiO_2)_{10}$–cholecalciferol (vitamin D3) predicted by DFT; selected values of bond distances in Å and bond energies in kcal mol^{-1}.

The possible formation of the $(TiO_2)_{10}$–cholecalciferol–glycerol structure involves different binding interactions in relation to the $(TiO_2)_{10}$–cholecalciferol formation mechanism. That difference is probably a consequence of higher steric as well as energetic requirements needed for the $(TiO_2)_{10}$–cholecalciferol–glycerol molecular interactions. These interactions require an additional step of releasing a few C–H···O hydrogen bonds in stable $(TiO_2)_{10}$–cholecalciferol structure (Figure 6) due of the access of one or more glycerol molecules. Namely, when the $(TiO_2)_{10}$–cholecalciferol–glycerol was formed, the cholecalciferol could have be attached to the $(TiO_2)_{10}$ cluster in different way related to $(TiO_2)_{10}$–cholecalciferol structure, as shown in Figures 6 and 7. It is important to note the large capacity of the cholecalciferol in the $(TiO_2)_{10}$–cholecalciferol structure to attract glycerol molecules due to stronger O–H···O hydrogen and weaker C–H···O hydrogen bonds ($E_{O...H}$ ranges from −0.82 kcal to −1.86 kcal mol^{-1}; $d_{O...H}$ ranges from 2.963 to 2.524 Å), as shown in Figure 7. However, the formation of the $(TiO_2)_{10}$-cholecalciferol-glycerol structure is slightly endergonic process (ΔG^*_{INT} = 5.51 kcal mol^{-1}) indicating that it could act only as a "coating intermediate", which is in a good agreement with FT-IR spectroscopy and contact angle measurement results (Figures 3 and 4). The most probably, as can be seen from the EIS response after one-day immersion of the modified Ti-implant (Figure 5b), this "coating intermediate" was responsible for temporal coating structural rearrangement that was reflected in the coating resistance value (R_c value, Table 3).

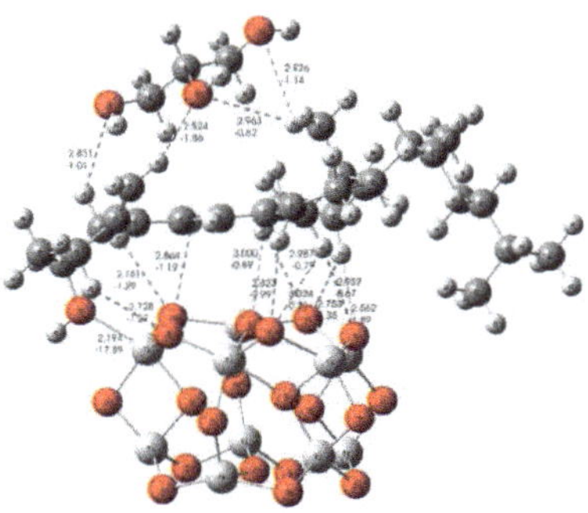

Figure 7. The most stable structure of the $(TiO_2)_{10}$–cholecalciferol (vitamin D3)-glycerol, predicted by DFT; selected values of bond distances in Å and bond energies in kcal mol^{-1}.

Based on experimental findings and DFT calculations, it seems that is not possible to predict the mechanism responsible for bioactive coating formation in the presence of solvent molecules solely based on the aspect of thermodynamic stability of the coating intermediate. For a more detailed/accurate coating formation mechanism, additional species present in the artificial saliva solution (simulated oral cavity physiological solution) should be taken into consideration due to their possible stronger coordination ability, which could replace the weaker hydrogen bonds between cholecalciferol and glycerol.

Taking into account the calculated energy values of common bonding (Ti–O) and eleven weaker C–H···O hydrogen bonds, as well as ΔG^*_{INT}, it can be concluded that cholecalciferol (D3 vitamin) molecules bond to the oxide-covered Ti implant surface displaying a high coverage level. Consequently, the bioactive coating on the Ti dental implant surface is very stable and resistant to the aggressive environment of artificial saliva, according to the EIS findings. Because of well-known bioactivity and, in this study, the proven protective effectiveness of D3 vitamin as a coating, a described strategy can be applied for a design of real dental implants of improved biocompatibility and osseointegrity.

4. Conclusions

In this study, the titanium dental implant was functionalized with bioactive molecules of vitamin D3 (cholecalciferol) to make its surface more corrosion resistant to aggressive oral cavity fluids and at the same time more attractive for bone cell adhesion. An integrated experimental–theoretical approach for the characterization of the Ti dental implant/bioactive coating interface was applied.

ATR-FTIR spectroscopy and contact angle measurement results confirmed the presence of the D3 layer onto the oxide covered Ti-implant surface.

Coating's formation mechanism was proposed, considering the DFT calculation results. D3 molecules possess an affinity to the titanium surface covered by the TiO_2 layer and are bonded to the surface through a Ti–O bond ($E_{Ti–O} = -14.68$ kcal mol^{-1}), which is stabilized by eleven weaker C–H···O hydrogen bonds ($E_{O\cdots H}$ ranges from −0.61 to −2.99 kcal mol^{-1}). This type of interactions resulted in a formation of the enclosed Ti-implant surface-vitamin D3 structure with high surface coverage level.

The bioactive coating influence on protective properties of the Ti-implant was investigated in situ by using EIS during different exposure times to an artificial saliva solution. The presence of the D3 layer on the oxide covered Ti-implant surface significantly changed the structure of the electrified Ti-implant surface/bioactive coating/artificial saliva interface compared to the Ti-implant surface/oxide film/artificial saliva interface.

During a prolonged immersion period, the stable bioactive coating effectively prevented the Ti-implant surface from contacting aggressive ions present in the artificial saliva solution. After a 7 day immersion period, the protection effectiveness amounted 97.5%.

Dental implant design that simultaneously stimulates an implant–bone connection creation (osseointegration) and enables good anticorrosion protection during an exposure to oral cavity fluids is desirable from a real application point of view.

Supplementary Materials: The following are available online at http://www.mdpi.com/2079-6412/9/10/612/s1, Computational modelling, Table S1: Formation of the most stable $(TiO_2)_{10}$−cholecalciferol[(a)], $(TiO_2)_{10}$−cholecalciferol−glycerol [(b)] and $(TiO_2)_{10}$−glycerol species[(c)]. Standard state (1M) free energies of interaction $\Delta_r G^*_{INT}$ computed by using the SMD solvation model at the M06/6-311++G(2df,2pd) + LANL2DZ//M06/6-31+G(d,p) + LANL2DZ level of theory, Table S2: Bond lengths (*d*), energies (*E*) and QTAIM properties of the selected bonds in the most stable $(TiO_2)_{10}$−cholecalciferol, $(TiO_2)_{10}$−cholecalciferol−glycerol and $(TiO_2)_{10}$−glycerol structures, Table S3: Total electronic energy, E^{Tot}_{soln}, obtained at the SMD/M06/6-311++G(2df,2pd) + LANL2DZ//SMD/M06/6-31+G(d,p) + LANL2DZ level of theory, thermal correction to the Gibbs free energy, $\Delta G^*_{VRT,soln}$, obtained at the SMD/M06/6-31+G(d,p) + LANL2DZ level of theory, and total free energy, G^*_X, ($G^*_X = E^{Tot}_{soln} + \Delta G^*_{VRT,soln}$) in 1,2-ethandiol media of the investigated species (all energies in hartree), Table S4: Optimized Cartesian coordinates of the calculated systems (given as XYZ format).

Author Contributions: Conceptualization, A.Š. and Ž.P.; Formal Analysis, J.K. and I.D.; Investigation, J.K., I.D., N.M. and Ž.P.; Methodology, J.K., A.Š., I.D. and Ž.P.; Resources, J.K., I.D., N.M. and M.P.; Supervision, A.Š. and Ž.P.; Writing–Original Draft, J.K., A.Š. and Ž.P.; Writing–Review and Editing, J.K., A.Š. and Ž.P.

Funding: This research received no external funding.

Acknowledgments: This work has been partially supported by SAFU, project KK.01.1.1.01.0001. The authors would like to thank the Zagreb University Computing Centre (SRCE) for generously granting computational resources on the ISABELLA cluster (isabella.srce.hr). The authors also thank Mirela Leskovac for her valuable comments in wettability measurements.

Conflicts of Interest: The authors declare no conflict of interest.

References

1. Eliaz, N. Corrosion of Metallic Biomaterials: A Review. *Materials* **2019**, *12*, 407. [CrossRef] [PubMed]
2. Hansen, D.C. Metal Corrosion in the Human Body: The Ultimate Bio-Corrosion Scenario. *Electrochem. Soc. Interface* **2008**, *17*, 31.
3. Gaviria, L.; Salcido, J.P.; Guda, T.; Ong, J.L. Current trends in dental implants. *J. Korean Assoc. Oral Maxillofac. Surg.* **2014**, *40*, 50–60. [CrossRef] [PubMed]
4. Yan, H.; Afroz, S.; Dalanon, J.; Goto, N.; Hosoki, M.; Matsuka, Y. Metal allergy patient treated by titanium implant denture: A case report with at least 4-year follow-up. *Clin. Case Rep.* **2018**, *6*, 1972–1977. [CrossRef] [PubMed]
5. Siddiqi, A.; Payne, A.G.T.; De Silva, R.K.; Duncan, W.J. Titanium allergy: Could it affect dental implant integration? *Clin. Oral Implant. Res.* **2011**, *22*, 673–680. [CrossRef] [PubMed]
6. Syed, M.; Chopra, R.; Sachdev, V. Allergic Reactions to Dental Materials—A Systematic Review. *J. Clin. Diagn. Res.* **2015**, *9*, ZE04. [CrossRef] [PubMed]
7. Sicilia, A.; Cuesta, S.; Coma, G.; Arregui, I.; Guisasola, C.; Ruiz, E.; Maestro, A. Titanium allergy in dental implant patients: A clinical study on 1500 consecutive patients. *Clin. Oral Implant. Res.* **2008**, *19*, 823–835. [CrossRef] [PubMed]
8. Smeets, R.; Stadlinger, B.; Schwarz, F.; Beck-Broichsitter, B.; Jung, O.; Precht, C.; Kloss, F.; Gröbe, A.; Heiland, M.; Ebker, T. Impact of Dental Implant Surface Modifications on Osseointegration. *BioMed Res. Int.* **2016**, *2016*, 6285620. [CrossRef]
9. Al Mugeiren, O.M.; Baseer, M.A. Dental Implant Bioactive Surface Modifiers: An Update. *J. Int. Soc. Prev. Community Dent.* **2019**, *9*, 1–4. [CrossRef] [PubMed]
10. Tian, B.; Xie, D.B.; Wang, F.H. Corrosion behavior of TiN and TiN/Ti composite films on Ti_6Al_4V alloy in Hanks solution. *J. Appl. Electrochem.* **2009**, *39*, 447–453. [CrossRef]
11. Yuan, H.; Yang, Z.; Li, Y.; Zhang, X.; De Bruijn, J.D.; De Groot, K. Osteoinduction by calcium phosphate biomaterials. *J. Mater. Sci. Mater. Electron.* **1998**, *9*, 723–726. [CrossRef]
12. Sousa, L.L.; Ricci, V.P.; Prado, D.G.; Apolinario, R.C.; Vercik, L.C.; Rigo, E.C.; Fernandes, M.C.; Mariano, N.A. Titanium Coating with Hydroxyapatite and Chitosan Doped with Silver Nitrate. *Mater. Res.* **2017**, *20*, 863–868. [CrossRef]
13. Łukaszewska-Kuska, M.; Krawczyk, P.; Martyla, A.; Hedzelek, W.; Dorocka-Bobkowska, B. Hydroxyapatite coating on titanium endosseous implants for improved osseointegration: Physical and chemical considerations. *Adv. Clin. Exp. Med.* **2018**, *27*, 1055–1059. [CrossRef] [PubMed]
14. Choi, A.H.; Ben-Nissan, B.; Matinlinna, J.P.; Conway, R.C. Current perspectives: Calcium phosphates nanocoatings and nanocomposite coatings in dentistry. *J. Dent. Res.* **2013**, *92*, 853–859. [CrossRef] [PubMed]
15. Rojo, L.; Gharibi, B.; McLister, R.; Meenan, B.J.; Deb, S. Self-assembled monolayers of alendronate on Ti_6Al_4V alloy surfaces enhance osteogenesis in mesenchymal stem cells. *Sci. Rep.* **2016**, *6*, 30548. [CrossRef] [PubMed]
16. Hu, X.; Neoh, K.G.; Shi, Z.; Kang, E.-T.; Wang, W. An In Vitro Assessment of Fibroblast and Osteoblast Response to Alendronate-Modified Titanium and the Potential for Decreasing Fibrous Encapsulation. *Tissue Eng. Part A* **2013**, *19*, 1919–1930. [CrossRef] [PubMed]
17. Trybek, G.; Aniko-Wlodarczyk, M.; Kwiatek, J.; Preuss, O.; Brodkiewicz, A.; Sinicyn, A.; Grzywacz, A. The effect of vitamin D3 on the osteointegration of dental implant. *Balt. J. Health Phys. Act.* **2018**, *10*, 25–33. [CrossRef]

18. Satué, M.; Monjo, M.; Ronold, H.J.; Lyngstadaas, S.P.; Ramis, J.M. Titanium implants coated with UV-irradiated vitamin D precursor and vitamin E: In vivo performance and coating stability. *Clin. Oral Implant. Res.* **2016**, *28*, 424–431. [CrossRef]
19. *ISO 5832-2:2018 Implants for surgery - Metallic materials—Part 2: Unalloyed titanium*; ISO: Geneva, Switzerland, 2018.
20. Available online: https://www.upmet.com/products/titanium/cp-grade-2 (accessed on 25 June 2019).
21. Gouzman, I.; Dubey, M.; Carolus, M.D.; Schwartz, J.; Bernasek, S.L. Monolayer vs. multilayer self-assembled alkylphosphonate films: X-ray photoelectron spectroscopy studies. *Surf. Sci.* **2006**, *600*, 773–781. [CrossRef]
22. Petrović, Ž.; Katić, J.; Metikoš-Huković, M.; Dadafarin, H.; Omanovic, S. Modification of a Nitinol Surface by Phosphonate Self-Assembled Monolayers. *J. Electrochem. Soc.* **2011**, *158*, F159–F165. [CrossRef]
23. Mellado-Valero, A.; Muñoz, A.I.; Pina, V.G.; Sola-Ruiz, M.F. Electrochemical Behaviour and Galvanic Effects of Titanium Implants Coupled to Metallic Suprastructures in Artificial Saliva. *Materials* **2018**, *11*, 171. [CrossRef] [PubMed]
24. Boukamp, A. A Nonlinear Least Squares Fit procedure for analysis of immittance data of electrochemical systems. *Solid State Ionics* **1986**, *20*, 31–44. [CrossRef]
25. Frisch, M.J.; Trucks, G.W.; Schlegel, H.B.; Scuseria, G.E.; Robb, M.A.; Cheeseman, J.R.; Scalmani, G.; Barone, V.; Mennucci, B.; Petersson, G.A.; et al. *Gaussian 09, Revision D.01*; Gaussian, Inc.: Wallingford, CT, USA, 2013.
26. Allard, M.M.; Merlos, S.N.; Springer, B.N.; Cooper, J.; Zhang, G.; Boskovic, D.S.; Kwon, S.R.; Nick, K.E.; Perry, C.C. Role of TiO_2 Anatase Surface Morphology on Organophosphorus Interfacial Chemistry. *J. Phys. Chem. C* **2018**, *122*, 29237–29248. [CrossRef]
27. Qu, Z.-W.; Kroes, G.-J. Theoretical Study of Stable, Defect-Free $(TiO_2)_n$ Nanoparticles with n = 10–16. *J. Phys. Chem. C* **2007**, *111*, 16808–16817. [CrossRef]
28. Zhao, Y.; Thrular, D.G. The M06 suite of density functionals for main group thermochemistry, thermochemical kinetics, noncovalent interactions, excited states, and transition elements: Two new functionals and systematic testing of four M06-class functionals and 12 other functionals. *Theor. Chem. Acc.* **2008**, *120*, 215–241. [CrossRef]
29. Zhao, Y.; Truhlar, D.G. Density Functionals with Broad Applicability in Chemistry. *Acc. Chem. Res.* **2008**, *41*, 157–167. [CrossRef] [PubMed]
30. Zhao, Y.; Truhlar, D.G. Density Functional Theory for Reaction Energies: Test of Meta and Hybrid Meta Functionals, Range-Separated Functionals, and Other High-Performance Functionals. *J. Chem. Theory Comput.* **2011**, *7*, 669–676. [CrossRef]
31. Wadt, W.R.; Hay, P.J. Ab initio effective core potentials for molecular calculations. Potentials for main group elements Na to Bi. *J. Chem. Phys.* **1985**, *82*, 284–298. [CrossRef]
32. Marenich, A.V.; Cramer, C.J.; Truhlar, D.G. Universal Solvation Model Based on Solute Electron Density and on a Continuum Model of the Solvent Defined by the Bulk Dielectric Constant and Atomic Surface Tensions. *J. Phys. Chem. B* **2009**, *113*, 6378–6396. [CrossRef] [PubMed]
33. Keith, T.A. AIMAll (Version 17.01.25). Available online: http://aim.tkgristmill.com/versionhistory.html (accessed on 12 February 2019).
34. Bader, R.F.W. *Atoms in Molecules: A Quantum Theory*; Oxford University Press: Oxford, NY, USA, 1994.
35. Rupp, F.; Scheideler, L.; Rehbein, D.; Axmann, D.; Geis-Gerstorfer, J. Roughness induced dynamic changes of wettability of acid etched titanium implant modifications. *Biomaterials* **2004**, *25*, 1429–1438. [CrossRef]
36. Webster, T.J.; Ejiofor, J.U. Increased osteoblast adhesion on nanophase metals: Ti, Ti6Al4V, and CoCrMo. *Biomatererials* **2004**, *25*, 4731–4739. [CrossRef] [PubMed]
37. Ferreira, C.C.; Pereira Ricci, V.; de Sousa, L.L.; Marianoa, N.A.; Campos, M.G.N. Improvement of Titanium Corrosion Resistance by Coating with Poly-Caprolactone and Poly-Caprolactone/Titanium Dioxide: Potential Application in Heart Valves. *Mater. Res.* **2017**, *20*, 126–133. [CrossRef]
38. Ingole, P.G.; Baig, M.I.; Choi, W.K.; Lee, H.K. Synthesis and characterization of polyamide/polyester thin-film nanocomposite membranes achieved by functionalized TiO_2 nanoparticles for water vapor separation. *J. Mater. Chem. A* **2016**, *4*, 5592–5604.
39. Othayoth, R.; Mathi, P.; Bheemanapally, K.; Kakarla, L.; Botlagunta, M. Characterization of vitamin–cisplatin-loaded chitosan nano-particles for chemoprevention and cancer fatigue. *J. Microencapsul.* **2015**, *32*, 578–588. [CrossRef] [PubMed]
40. Orazem, M.E.; Tribollet, B. *Electrochemical Impedance Spectroscopy*; Wiley: Hoboken, NJ, USA, 2008; pp. 233–265.

41. Brug, G.; Eeden, A.V.D.; Sluyters-Rehbach, M.; Sluyters, J. The analysis of electrode impedances complicated by the presence of a constant phase element. *J. Electroanal. Chem. Interfacial Electrochem.* **1984**, *176*, 275–295. [CrossRef]
42. Jorcin, J.-B.; Orazem, M.E.; Pébère, N.; Tribollet, B. CPE analysis by local electrochemical impedance spectroscopy. *Electrochim. Acta* **2006**, *51*, 1473–1479. [CrossRef]
43. Pan, J.; Thierry, D.; Leygraf, C. Electrochemical impedance spectroscopy study of the passive oxide film on titanium for implant application. *Electrochimica Acta* **1996**, *41*, 1143–1153. [CrossRef]
44. Kosec, T.; Legat, A.; Kovač, J.; Klobčar, D. Influence of Laser Colour Marking on the Corrosion Properties of Low Alloyed Ti. *Coatings* **2019**, *9*, 375. [CrossRef]
45. Aziz-Kerrzo, M.; Conroy, K.G.; Fenelon, A.M.; Farrell, S.T.; Breslin, C. Electrochemical studies on the stability and corrosion resistance of titanium-based implant materials. *Biomaterials* **2001**, *22*, 1531–1539. [CrossRef] [PubMed]
46. De Assis, S.L.; Wolynec, S.; Costa, I. Corrosion characterization of titanium alloys by electrochemical techniques. *Electrochim. Acta* **2006**, *51*, 1815–1819. [CrossRef]
47. Qu, Q.; Wang, L.; Chen, Y.; Li, L.; He, Y.; Ding, Z. Corrosion Behavior of Titanium in Artificial Saliva by Lactic Acid. *Materials* **2014**, *7*, 5528–5542. [CrossRef] [PubMed]
48. Guo, X. Corrosion and Electrochemical Impedance Properties of Ti Alloys as Orthopaedic Trauma Implant Materials. *Int. J. Electrochem. Sci.* **2017**, *12*, 9007–9016. [CrossRef]
49. Mishnaevsky, L.; Levashov, E.; Valiev, R.Z.; Segurado, J.; Sabirov, I.; Enikeev, N.; Prokoshkin, S.; Solov'Yov, A.V.; Korotitskiy, A.; Gutmanas, E.; et al. Nanostructured titanium-based materials for medical implants: Modeling and development. *Mater. Sci. Eng. R Rep.* **2014**, *81*, 1–19. [CrossRef]
50. Scully, J.R. Polarization Resistance Method for Determination of Instantaneous Corrosion Rates. *Corrosion* **2000**, *56*, 199–218. [CrossRef]
51. Tamilselvi, S.; Murugaraj, R.; Rajendran, N. Electrochemical impedance spectroscopic studies of titanium and its alloys in saline medium. *Mater. Corros.* **2007**, *58*, 113–120. [CrossRef]
52. Evans, S.D.; Urankar, E.; Ulman, A.; Ferris, N. Self-assembled monolayers of alkanethiols containing a polar aromatic group: Effects of the dipole position on molecular packing, orientation, and surface wetting properties. *J. Am. Chem. Soc.* **1991**, *113*, 4121–4131. [CrossRef]
53. Boubour, E.; Lennox, R.B. Insulating Properties of Self-Assembled Monolayers Monitored by Impedance Spectroscopy. *Langmuir* **2000**, *16*, 4222–4228. [CrossRef]

MDPI AG
Grosspeteranlage 5
4052 Basel
Switzerland
Tel.: +41 61 683 77 34

Coatings Editorial Office
E-mail: coatings@mdpi.com
www.mdpi.com/journal/coatings

www.ingramcontent.com/pod-product-compliance
Lightning Source LLC
Chambersburg PA
CBHW061622210926
51850CB00009B/2409
9783725846559